Sharing the Road

With the advancements in sensor and computation technologies, connected autonomous vehicle (CAV) services are being tested and deployed on public roads, sharing the road with human-driven vehicles (HVs). This brings challenges to traffic safety, due to uncertainty in HVs behaviors. To address these challenges, *Sharing the Road* offers solutions to create a safer world where HVs and AVs mix freely on our roads.

Presented from a team of researchers from different domains offering varying perspectives, this book systematically summarizes the issues and trends of mixed traffic, from the micro (such as vehicle algorithm design) to the macro (including traffic flow optimization and policy) perspectives. Customized car-following models describing the behaviors of HVs when interacting with CAVs and control algorithms optimizing the traffic flow in terms of traffic safety and efficiency are discussed. The title also summarizes approaches to model HVs' behaviors and case studies of real-life examples are provided, leaving the reader with a detailed vision of the future of a mixed-traffic world.

This title will appeal to professionals at any career stage in the fields of human factor engineering, traffic engineering, urban design, and vehicle engineering.

Sharing the Road

Understanding and Facilitating Mixed Traffic with Human-Driven and Autonomous Vehicles

Edited by
Dengbo He, Xiaotong Sun, and Huan Yu

CRC Press
Taylor & Francis Group
Boca Raton London New York

CRC Press is an imprint of the
Taylor & Francis Group, an **informa** business

Designed cover image: image credited to DedMityay [ShutterStock ID: 1739926457]

First edition published 2025
by CRC Press
2385 NW Executive Center Drive, Suite 320, Boca Raton FL 33431

and by CRC Press
4 Park Square, Milton Park, Abingdon, Oxon, OX14 4RN

CRC Press is an imprint of Taylor & Francis Group, LLC

ISBN: 9781032639406 (hbk)
ISBN: 9781032674681 (pbk)
ISBN: 9781032674667 (ebk)

DOI: 10.1201/9781032674667

Typeset in Times
by Newgen Publishing UK

Contents

Acknowledgements

This work was supported by the National Natural Science Foundation of China (no. 52202425) and the Nansha District Key Area S&T Scheme, Science and Technology Bureau of Nansha District (no. 2023ZD006).

About the Editors

Dengbo He received his Bachelor's degree from Hunan University in 2012, MS degree from Shanghai Jiao Tong University in 2016, and PhD degree from the University of Toronto in 2020. He worked as a postdoc research fellow at the University of Toronto from 2020 to 2021 and is currently an Assistant Professor in Intelligent Transportation Thrust and Robotics and Autonomous Systems Thrust, at the Hong Kong University of Science and Technology (HKUST) (Guangzhou). His research focuses on human factors, driver behavior, and the interactions between driving automation and human drivers.

Xiaotong Sun is an assistant professor in the Intelligent Transportation Thrust and an affiliated assistant professor in the Civil and Environmental Engineering Department at the Hong Kong University of Science and Technology. Before joining HKUST, she was a postdoc research fellow at the University of Michigan, Ann Arbor. She received her PhD in Civil Engineering from the University of Michigan in July 2020.

Huan Yu is an Assistant Professor in the Intelligent Transportation Thrust of the Systems Hub, and an affiliated Assistant Professor in the Department of Civil and Environmental Engineering at the Hong Kong University of Science and Technology (HKUST). Yu received her BEng degree in Aerospace Engineering from Honor School (Elite program) of Northwestern Polytechnical University, and the MSc and PhD degrees in Aerospace Engineering from the Department of Mechanical and Aerospace Engineering, University of California, San Diego. She was a visiting scholar at University of California, Berkeley in 2018 and Massachusetts Institute of Technology in 2019. She was a postdoc researcher at University of California, San Diego before joining the HKUST (Guangzhou) in 2021.

Contributors

Dengbo He
Thrust of Intelligent Transportation
The Hong Kong University of Science and Technology (Guangzhou)
Guangzhou, China

Chunxi Huang
School of Design
Hunan University
Changsha, China
and
Thrust of Robotics and Autonomous Systems
The Hong Kong University of Science and Technology (Guangzhou)
Guangzhou, China

Qinhua Jiang
Department of Civil and Environmenta Engineering, UCLA Mobility Lab
University of California at Los Angeles
Los Angeles, CA, USA

Tianhao Li
Thrust of Intelligent Transportation
The Hong Kong University of Science and Technology (Guangzhou)
Guangzhou, China

Yang Liu
Thrust of Intelligent Transportation
The Hong Kong University of Science and Technology (Guangzhou)
Guangzhou, China

Jiaqi Ma
Department of Civil and Environmental Engineering, UCLA Mobility Lab
University of California at Los Angeles
Los Angeles, CA, USA

Jian Rong
School of Civil Engineering and Transportation
Guangzhou University
Guangzhou, China

Xiaotong Sun
Thrust of Intelligent Transportation
The Hong Kong University of Science and Technology (Guangzhou)
Guangzhou, China

Ran Tu
School of Transportation
Southeast University
Nanjing, China

Yi Wang
School of Civil Engineering and Transportation
Guangzhou University
Guangzhou, China

Shan Xue
School of Transportation
Southeast University
Nanjing, China

Huan Yu
Thrust of Intelligent Transportation
The Hong Kong University of Science and Technology (Guangzhou)
Guangzhou, China

Chen Yue
School of Civil Engineering and Transportation
Guangzhou University
Guangzhou, China

Brian Yueshuai He
Department of Civil and Environmental Engineering, UCLA Mobility Lab
University of California at Los Angeles
Los Angeles, CA, USA

Jiahao Zhang
Thrust of Intelligent Transportation
The Hong Kong University of Science and Technology (Guangzhou)
Guangzhou, China

Chenjing Zhou
School of Civil Engineering and Transportation
Guangzhou University
Guangzhou, China

Xiaoyu Zhou
Thrust of Intelligent Transportation
The Hong Kong University of Science and Technology (Guangzhou)
Guangzhou, China

Meixin Zhu
Thrust of Intelligent Transportation
The Hong Kong University of Science and Technology (Guangzhou)
Guangzhou, China

1 Introduction

Dengbo He, Huan Yu, and Xiaotong Sun

1.1 WHAT IS MIXED TRAFFIC

Traditionally, the term "mixed traffic" refers to the heterogeneous traffic flow containing various vehicles—either motorized or non-motorized (Hidayati et al., 2014). In recent years, with the advancement of autonomous driving technologies and communication technologies, fully autonomous vehicles (AVs) or vehicles with some level of driving automation have been gradually entering the market. However, before they saturate the market, AVs and human-driven vehicles (HDVs) will inevitably have to share the road, leading to the traffic that consists of both AVs and HDVs. Being different from the traditional mixed traffic that consists of human road agents, the coexistence of both AVs and HDVs on the road can be more complex, given the subtle relationships between human users and automation (e.g., Madhavan & Wiegmann, 2007). In the current book, we will mainly focus on the traffic stream of AVs and HDVs, and frame such traffic stream as **mixed traffic** throughout the book.

1.2 THE RATIONALE OF THIS BOOK

Introducing AVs into the traffic flow brings several challenges and opportunities to traffic management and control. Although researchers in past decades have explored a wide range of traffic modeling and control topics, including the impact of new technologies (e.g., artificial intelligence, Hu et al., 2023), the introduction of AVs in the traffic flow can be more dramatic. Some key points are outlined here below:

- **AVs are still less than ideal and have different characteristics compared to HDVs.** Although driving automation technologies have been evolving dramatically and the crash rates of AVs have been decreasing in recent years (Zhou et al., 2024), in the foreseeable future, driving automation may perform differently compared to human drivers. Thus, the characteristics of AVs can impact traffic flow and traffic management. However, the AV-related data remains sparse at this stage. Hence, on the one hand, it is necessary to review the approaches taken by existing research that focused on mixed traffic and inform future research directions. On the other hand, it is also necessary to summarize the characteristics of the AVs in mixed traffic to inform future research.

DOI: 10.1201/9781032674667-1

- **The human drivers' adaptation to AVs needs to be considered.** It has been widely acknowledged that human operators may adapt their behaviors when interacting or cooperating with automation (Lee & Moray, 1994). Human road users may exhibit different strategies when interacting with AVs versus interacting with other HDVs. Hence, when modeling the mixed traffic, it is necessary to consider the behavioral changes of the HDVs. This book will review some of the approaches that have been found to be effective in modeling behavioral changes and discuss some of the behavioral changes that have already been recognized among human road users in mixed traffic.
- **Human road agents' attitudes and acceptance of AVs matter to mixed traffic.** Although AVs can be fully rational and designed to modulate traffic for specific purposes, the realization of these benefits depends on the design of AVs, the users' or potential users' acceptance/attitudes toward the AVs, and the correct modeling of the human-AV interactions. In this case, the research in mixed traffic also needs to take the road users' heterogeneity into consideration and model such heterogeneity when designing the policies or traffic control strategies. This is a relatively new perspective for traffic research, and we will also briefly discuss the approaches and findings in this direction.

1.3 CHAPTERS OVERVIEW

In **Chapter 2**, we will briefly review the empirical research approaches that have been used or can be used in mixed traffic research for data collection. The pros and cons of different research approaches will be discussed.

In **Chapter 3**, we will mainly discuss the tools for microscopic mixed traffic simulation, including data-driven and model-based approaches.

In **Chapter 4**, instead of focusing on micro driver behaviors, we provide more details regarding the traffic flow modeling in mixed traffic.

Chapter 5 will further zoom out and compare macroscopic and microscopic traffic flow models in artificial and mixed traffic scenarios, highlighting their differences and applications. It also discusses simulation tools like VISSIM (www.ptvgroup.com/en/products/ptv-vissim), AIMSUN (www.aimsun.com), and SUMO (https://eclipse.dev/sumo/), which are used to model and evaluate these traffic conditions.

Chapter 6 will discuss more about the characteristics of mixed traffic from the traffic safety perspective of view and **Chapter 7** will discuss the mixed traffic from the environmental impact perspective of view.

Finally, we provide three case studies, covering the topic of driver behavior modeling in mixed traffic (**Case Study 1**), modeling of mixed traffic flow (**Case Study 2**), and users' acceptance of mixed traffic as well as its impact on vehicle miles traveled and emissions (**Case Study 3**).

REFERENCES

Hidayati, N., Montgomery, F., & Liu, R. (2014). *Indonesian Traffic Conditions in Context Kondisi Lalu Lintas Indonesia dalam Konteks*. http://publikasiilmiah.ums.ac.id/handle/11617/5174

Hu, W. X., Ishihara, H., Chen, C., Shalaby, A., & Abdulhai, B. (2023). Deep Reinforcement Learning Two-Way Transit Signal Priority Algorithm for Optimizing Headway Adherence and Speed. *IEEE Transactions on Intelligent Transportation Systems*, *24*(8), 7920–7931. https://doi.org/10.1109/TITS.2023.3266461

Lee, J. D., & Moray, N. (1994). Trust, self-confidence, and operators' adaptation to automation. *International Journal of Human-Computer Studies*, *40*(1), 153–184. https://doi.org/10.1006/ijhc.1994.1007

Madhavan, P., & Wiegmann, D. A. (2007). Similaritics and differences between human–human and human–automation trust: An integrative review. *Theoretical Issues in Ergonomics Science*. https://doi.org/10.1080/14639220500337708

Zhou, R., Zhang, G., Huang, H., Wei, Z., Zhou, H., Jin, J., Chang, F., & Chen, J. (2024). How would autonomous vehicles behave in real-world crash scenarios? *Accident Analysis & Prevention*, *202*, 107572. https://doi.org/10.1016/j.aap.2024.107572

2 Empirical Research for Mixed Traffic Research

Dengbo He, Xiaoyu Zhou, Yang Liu, and Jiahao Zhang

2.1 WHY EMPIRICAL RESEARCH

The survey-based approach has been widely adopted for mixed traffic research in the past few years. It has shown a great advantage in understanding users' concerns and acceptance of autonomous driving before the technology matures. For example, T. Zhang et al. (2020) explored the factors influencing users' acceptance of autonomous vehicles (AVs) through a survey study that included 647 Chinese drivers. Similarly, based on a survey study with 451 valid samples, Wu et al. (2023) found that the level of automation can directly or indirectly (through trust) affect public acceptance of AV, while demographic characteristics of respondents can only indirectly affect public acceptance of AV through trust. However, the survey data can hardly facilitate an understanding of the micro and meso behaviors for several reasons. First, up until now, fully AVs have not been officially commercialized, and a few people have experienced or even witnessed fully AVs running on public roads. Thus, their attitudes towards AVs were mostly based on their limited knowledge about AVs, and thus, their attitudes might be biased in such situations (Janatabadi & Ermagun, 2022). Second, data-driven, learning-based, or even model-based research all rely on users' micro and meso behaviors in mixed traffic, which cannot be captured in survey studies. Thus, collecting empirical data in mixed traffic is vital for the optimization of AV control algorithms and mixed traffic management. In this Chapter, we will mainly review the research that adopted empirical approaches, that is, an approach that gains knowledge or obtains empirical evidence through direct and indirect observation or experience. Specifically, we will discuss four commonly used approaches in driving behavioral research, including driving simulation research, field studies, naturalistic driving research, and observational research. We will briefly discuss the pros and cons of the approaches, introduce the basic procedures for conducting research with these approaches, and discuss the findings generated from research adopting these research methods.

DOI: 10.1201/9781032674667-2

2.2 TYPES OF EMPIRICAL DATA COLLECTION METHODS ADOPTED FOR MIXED TRAFFIC RESEARCH

2.2.1 Driving Simulation Approach

2.2.1.1 Driving Simulator Studies

The driving simulation is the most widely adopted approach in the driving behavior research domain. In driving simulator studies for mixed traffic, experimenters can design the traffic scenarios and have participants act as either drivers or other road agents (e.g., pedestrians or cyclists) in the scenarios. Given that AVs can be easily simulated in the simulator, drivers' responses to the AVs in mixed traffic can be readily captured at a relatively low cost in a well-controlled environment. For example, with the driving simulator, all participants in a study can experience the same scenarios, and the experimenters can easily control the order of the experimental conditions, which is especially beneficial for isolating the impacts from the factors of interest. The highly risky scenarios or tasks can also be tested in driving simulators. However, the pros of the driving simulator may also bring cons. For example, to isolate the influential factors in a driving simulator experiment, the complexity and, thus, the reality of the scenarios might be compromised. The lack of real risk and adverse consequences in driving simulator experiments may also bias users' behaviors and raise questions on the validity of the experiment outcomes, especially when quantitative data is needed.

2.2.1.2 Types of Driving Simulators

The validity of the driving simulator studies highly depends on the fidelity of the simulator. Although no strict standards define the level of fidelity of a driving simulator, previous research usually framed their simulators as low-, medium-, and high-fidelity ones. In general, this categorization is based on the hardware of the system, including the degrees of freedom (DoF), the cabin fidelity, and the display systems. For example, a computer system without motion functions (i.e., fixed-base) and non-cabin setup is usually considered a low-fidelity one (Figure 2.1a). The fixed system, or the system with low DoF (i.e., ≤ 3) and with a cabin-like setup, can be considered as a medium one (Figure 2.1b). While the high-fidelity simulators usually have a high DoF (≥ 6) motion platform, over 180-degree view angle, and at least a quarter-cabin (Figure 2.1c). However, the fidelity, by definition, should be more than hardware-related and can also be related to the simulation software. For example, a train simulator study (Olsson, 2023) found that the driving performance of train drivers in physically low-fidelity but functionally high-fidelity simulators is comparable to real train driving performance. Functional fidelity, however, is rarely mentioned in the driving research, potentially because functions related to the specific research purposes usually work well and hence can be regarded as with high fidelity in most cases, while the functions non-related to the specific research are either well-controlled (e.g., some driving simulators cannot turn the steering when lane centering control works) or not being used in the research. Table 2.1 summarizes the description of the simulators in some of the mixed-traffic-related research.

TABLE 2.1
Descriptions of the simulators used in mixed-traffic-related research up to 2023

Research	Description of the simulator	Self-identified level of fidelity
Manual drivers' experience and driving behavior in repeated interactions with automated Level 3 vehicles in mixed traffic on the highway (Stange et al., 2022a)	"The driving simulation was programmed using SILAB 6.0 and consists of a physical mock-up seat box with driver and passenger seats. The three LCD projectors (1920 × 1080 pixels each) projected the scenery onto an array of three screens (2 m × 2 m each) covering a field of view of 180°"	Medium-fidelity simulator
Safety at first sight?—Manual drivers' experience and driving behavior at first contact with Level 3 vehicles in mixed traffic on the highway (Stange et al., 2022b)		
The impact of expectations about automated and manual vehicles on drivers' behavior: Insights from a mixed traffic driving simulator study (Miller et al., 2022)	"Consisted of a vehicle mockup with an adjustable seat, FanaTech steering wheel, and pedals with force feedback…Run with SILAB 6.5 on an RTX3070 PC and three 55-inch 4K displays with a field of view of about 160 degrees"	Not mentioned
Driver-automated vehicle interaction in mixed traffic: Types of interaction and drivers' driving Styles (Ma & Zhang, 2024)	"Each simulator consisted of a vehicle mockup with an adjustable seat, FanaTech steering wheel, and pedals with force feedback ... The simulation was run with SILAB 6.5 on an RTX3070 PC and three 55-inch 4K displays with a field of view of about 160 degrees."	Not mentioned
Effects of marking automated vehicles on human drivers on highways (Fuest et al., 2020)	"The basis of the static driving simulator was a BMW 6 series mockup. A 6-channel projection system provided a realistic driving environment, with a refresh rate of 60 Hz. Three projectors were used for the 180° front view, and three projectors for the rear view (side and rear mirrors). We used the driving simulation software SILAB 6.5 of the Würzburg Institute for Traffic Sciences GmbH [29] and logged the driving data with 240 Hz. A 6-channel noise simulation completed the	Medium-fidelity simulator

TABLE 2.1 (Continued)
Descriptions of the simulators used in mixed-traffic-related research up to 2023

Research	Description of the simulator	Self-identified level of fidelity
	driving simulation. A freely programmable instrument cluster was used as human–machine interface. A tachometer and a speedometer were implemented for displaying driving-relevant information in this study."	
Risk modeling and quantification of a platoon in mixed traffic based on the mass-spring-damper model (Jiang et al., 2020)	"Mainly includes a Logitech G29 vehicle controller (steering wheel, pedal, and gear lever) and two sorts of simulation software (PreScan and Matlab/Simulink). "	Not mentioned
Driver behavior at a freeway merge to mixed traffic of conventional and connected autonomous vehicles (Chityala et al., 2020)	"DriveSafety Model CDS-250 Driving Simulato. the drivers sit in a partial cab based on a Ford Focus sedan. the simulator consists of various computing units, with most on the automotive frame containing the sedan, and a separate station for authoring scenarios for the simulations."	Not mentioned
Effect of cognitive distraction on physiological measures and driving performance in traditional and mixed traffic environments (Hua et al., 2021)	"The driving environment was established using the simulation software UC-win/Road and displayed on three 32-in LED displays. The horizontal viewing angle of the scenario display system was 120°, The sampling frequency was 100 Hz."	Low-fidelity simulator
Learning in mixed traffic: Drivers' adaptationtoambiguous communication depending on their expectations toward automated and manual vehicles (Miller et al., 2023)	"a vehicle mockup featuring an adjustable seat, force-feedback racing wheel (FanaTec Base V2), and pedals (Fanatec CSL Elite). 17-inch touchscreen mounted in the position of the vehicle's center console. Vehicle's front view was simulated on three 55-inch 4K UHD LED TVs, which were placed in front of the mockup, and created a driver's view of around 120 degrees."	Medium-fidelity simulator

(*continued*)

TABLE 2.1 (Continued)
Descriptions of the simulators used in mixed-traffic-related research up to 2023

Research	Description of the simulator	Self-identified level of fidelity
Cooperative driving in mixed traffic of manned and unmanned vehicles based on human driving behavior understanding (J. Lu et al., 2023)	"The driving simulator has 3 connected monitors for in-vehicle view, an extra monitor for control and data display, and a Logitech G290 driving force suit (a steering wheel, pedals, and a shifter)."	Not mentioned
Development of a research testbed for cooperative driving in mixed traffic of human-driven and autonomous vehicles (J. Lu et al., 2022)		
The effect of visual advanced driver assistance systems on a following human driver in a mixed-traffic condition (Ariansyah et al., 2023)	"The simulator consisted of three screens combined providing 36 degrees and 175 degrees on the vertical and horizontal field of view (FoV), respectively. It was also equipped with the steering wheel, pedals, and adjustable driver seat, while the control of the vehicle was implemented with automatic transmission."	Low-fidelity simulator

In addition to research with a single driving simulator, in recent years, to better understand users' behaviors in mixed traffic, researchers also started to conduct research with multiple driving simulators or the combination of driving simulators and walking simulators. For example, Kalantari et al. (2023) replicated vehicle–pedestrian interaction in a safe and controllable virtual environment (32 pairs, one driver and one pedestrian interacting with each other in different scenarios). The results show that kinematic cues play a greater role than psychological characteristics, such as sensory seeking and social value orientation, in determining who passes an unmarked intersection first. In Miller et al. (2022), multi-agent simulators were also used to investigate how the role of expectations shapes drivers' reactions to AVs and human-driven vehicles and found that human driving behavior can be improved when AVs match one's expectations. These types of research can provide insights into the more complex gaming processes in human–AV or AV–pedestrian interactions.

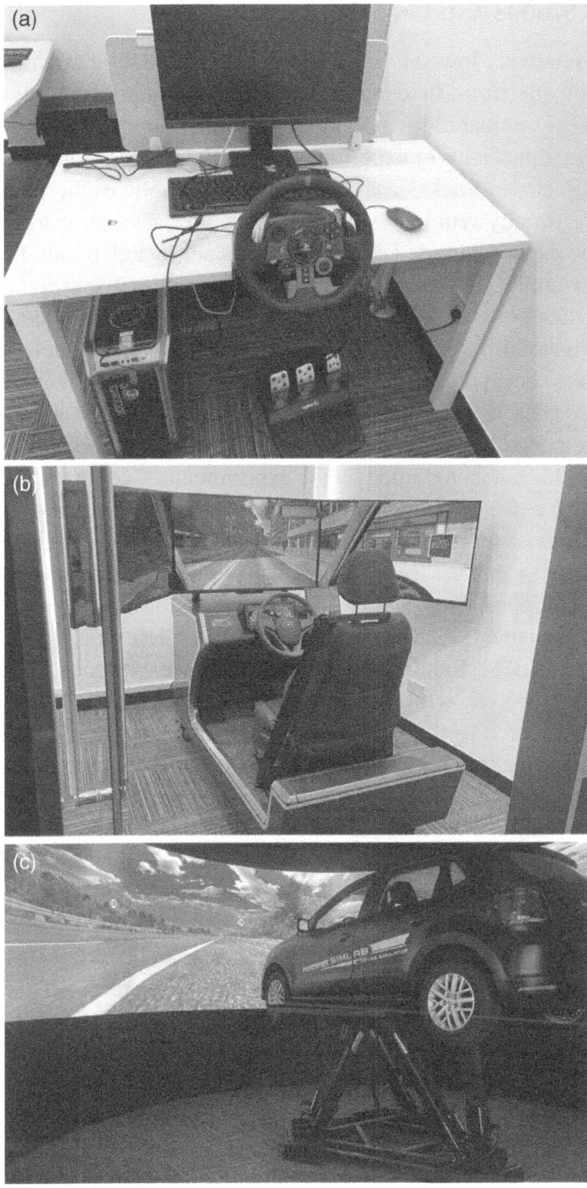

FIGURE 2.1 Driving simulation systems at the Hong Kong University of Science and Technology (Guangzhou): (a) low-fidelity simulator; (b) medium-fidelity simulator; (c) high-fidelity simulator.

2.2.2 Field Studies and On-Road Studies

Given that the relatively low validity of the driving simulation might be a concern for specific research questions, field studies can be an alternative option, which has relatively low risk (as compared to experiments on open public roads) and high validity but still reserves some levels of experiment controllability. In general, the field studies are conducted in closed tracks or test fields without public traffic. In the experiment, participants are usually required to conduct specific tasks in instrumented vehicles provided by the experimenters. Different sensors are installed onto the instrumented vehicles, for example, eye-tracking devices, physiological sensors, radars, and onboard vehicle data collection devices. A. Zhou et al. (2023) collected eye-tracking data to investigate different connected and autonomous vehicle (CAV) control settings on users' acceptance of CAV and their distraction behaviors. However, it should be noted that, as the experiment was still conducted in a contrived environment, the participants' behaviors may still be skewed to some extent. Specifically, the participants would be accompanied by the experimenters, and irrelevant traffic would be removed. Hence, the participants would experience lower than usual risks.

To simulate the risk more realistically, researchers may also conduct on-road studies using instrumented vehicles. Being different from field studies, the participants may drive vehicles equipped with data collection devices on public roads while accompanied by experimenters. As can be imagined, the fidelity of the scenarios will be higher compared to field studies, and participants may experience higher levels of risk on public roads. However, with such an approach, the controllability of the scenarios will be lower compared to the field experiments, as although the experimenters can control some static factors related to the scenario (e.g., the time of day, the weather, and the experimental areas), they have little to no control over the traffic scenarios surrounding the instrument vehicles.

At the same time, given that fully AVs are not mature enough to understand drivers' or pedestrians' responses to the AVs, the Wizard-of-Oz (Dahlbäck et al., 1993) technique was commonly adopted. With this approach, the system was simulated (e.g., by having an experimenter driving a non-autonomous vehicle), but the participants were not informed of such setup. Hence, participants may perceive the vehicle as fully autonomous. With this approach, previous researchers have successfully simulated the operation of AVs on the road. For example, Detjen et al. (2020) used the Wizard-of-Oz method to simulate an autonomous driving system. All participants were told that the vehicle was controlled by the driving automation, but it was actually controlled by an experimenter. Users' feedback and behavioral data were collected. Given that some (4 out of 12) participants perceived the system as human-operated, the authors compared how the trust in the AV varied across the two groups of participants, and it was found that participants trusted the human driver more.

2.2.3 Naturalistic Driving Research

The naturalistic driving study is a powerful tool for obtaining realistic driving behaviors among drivers. Being different from field studies, sensors and equipment are installed in a vehicle that can run on public roads. The vehicle can either be the participants' own vehicle (Liu et al., 2023) or the vehicle that is rented to the participants (e.g., Advanced Vehicle Technology Consortium; Hong et al., 2021). With such a setup, in naturalistic driving studies, the experimenters can observe realistic driving behaviors with minimum interference from the experimenters or the context of the research. However, the disadvantage is also obvious. Given that the study is usually conducted on public roads, the experimenters had little to no control over the traffic scenarios and states of the drivers. Thus, the density of the meaningful data points for a specific research question might be sparse, and data cleaning would be time-consuming. This type of research also has high requirements for data storage and transfer. For example, the largest and the most commonly used naturalistic driving dataset, the SHRP2, contains over 2 PB data for nearly 2,360 participants, leading to 3,700 participant-years data in total (Hallmark et al., 2015). Table 2.2 summarizes some of the existing public datasets from naturalistic driving studies.

Another major limitation of the naturalistic driving approach when applied to mixed traffic research is that, so far, existing naturalistic driving research has all focused on drivers' behaviors in the ego-vehicle. Although some research has used vehicles with some levels of driving automation (Wen et al., 2023), they provide limited knowledge on how human drivers respond to other AVs or vehicles controlled by driving automation in mixed traffic. Such kind of research is limited by two constraints at this stage. On the one hand, to collect the interactions with other road agents on the road, the ego-vehicle will need to be equipped with appropriate external sensors (e.g., LiDAR (Light Detection and Ranging) and cameras), which are expensive and may lead to privacy concerns. On the other hand, even when the instrumented vehicles with necessary sensors are available, collecting data regarding mixed traffic is still difficult, given that only a small portion of vehicles are equipped with advanced driving assistant systems (ADASs) or autonomous driving systems. However, it should be pointed out that knowing how human drivers respond to automated road agents and evaluate users' cognitive and decision-making processes would facilitate more efficient and explainable AV control algorithms. With more AVs being available on public roads, future naturalistic driving research may better capture human–AV interactions from an ego-vehicle-driver behavioral perspective of view.

2.2.4 Observational Approach

Though capturing human–AV interactions through a naturalistic driving research approach is difficult due to the sparsity of the AVs on public roads at this stage, observational studies can be a feasible option at this stage and have been widely adopted in previous research, given that certain cities or areas have allowed the AVs to test and operate commercially on public roads. Depending on the resolution of the datasets, the observational datasets can be categorized into two types: the trajectory-oriented one and the safety-oriented one. The former focuses more on the micro trajectories of human–AV interactions on public roads, while the latter focuses more on the safety records of mixed traffic.

In traditional traffic research, the trajectory-oriented data can be collected through roadside units (e.g., U.S. Department of Transportation Federal Highway Administration, n.d.), on-road data collection vehicles (e.g., K. Chen et al., 2024),

TABLE 2.2
Available public naturalistic driving studies up to 2023

Dataset	Vehicle type	Total length of data duration	Locations of data collection	Types of collected data	Research
UDRIVE	Cars, trucks, and powered two-wheelers	88,000 hours of vehicle data	Six European Union's Member States	Video and vehicle Sensors data	(Eenink et al., 2014)
CNDS	Conventional vehicle	Over 53,000 vehicle hours	Canada	Video of driver, video and vehicle sensors data	(Klauer et al., 2018)
ANDS	Private vehicle	360 volunteer drivers for 4 months	NSW and Victoria of Australian	Video of driver and other road users' behavior, video and sensors data of vehicle	(Williamson et al., 2015)
IVBSS	Light vehicles	Over 213,000 miles accumulated	Parts of southeast Michigan, US	Video and sensors data of drivers and vehicles; Assessment of drivers	(Steve, 2011)
SHNDS	Cars	Over 5,000 hours of continuous driving data from 60 drivers	Shanghai city of China	Video of driver, video and sensors data of vehicle, assessment	(Guo et al., 2022)
NTDS	Private vehicle	Nearly 102,000 trips over 800,000 km (500,000 miles).	Commonwealth of Virginia in the United States	Video of driver, video and sensors data of vehicle	(Lee et al., 2011)
SHRP2	Conventional vehicle	3.700 participant-years	Several cities in US	Video and vehicle sensors data	(National Research Council (U.S.), 2001)

and unmanned aerial vehicles (e.g., Krajewski et al., 2018). The Next Generation Simulation (NGSIM) Open Data used cameras mounted on highways to collect vehicle trajectories on highways. However, given the sparsity of the AVs on the road, collecting mixed traffic data with fixed sensors or cameras is inefficient at this stage. Thus, they may not be suitable for mixed traffic research until AVs start to saturate the market.

To date, existing trajectory-oriented mixed traffic datasets were mostly collected using AV-mounted sensors. The three most representative ones are the Waymo Open Dataset, the Lyft Level-5 Dataset, and the nuScenes dataset. All of them provided perceived speed and location information of the road agents surrounding the AVs at a relatively high sampling frequency (e.g., 10 Hz in Waymo Open Dataset). Because of the perception capabilities of the AVs, the datasets contain information regarding the road agents that are hundreds of meters away from the AV. Thus, representative traffic scenarios such as car-following behaviors and lane-changing behaviors can be extracted. Specifically, the Waymo Open Dataset consists of two parts: perception and motion. The perception part contains 1,000 20-second video clips, each of which is composed of well-synchronized and calibrated high-resolution LiDAR and camera data recorded in urban and suburban areas. The motion part consists of 103,354 20-second video clips representing 574 hours of driving data collected over 1,750 km of roadways. The Lyft Level-5 Dataset contains information collected by seven cameras and three LiDARs on a fleet of 25 SAE Level-5 AVs operated by Lyft, which includes map information covering more than 4,000 roads, 197 crosswalks, 60 stop signs, and 54 parking areas. Visible traffic participants, including vehicles, pedestrians, and cyclists within each scene, were also detected, and their motions, such as velocity, acceleration, and yaw rate, were provided. In total, the dataset also includes more than 55,000 3D human-annotated frames. The nuScenes dataset was collected by AVs equipped with LiDARs, radars, and cameras in four cities around the world, and it contains 1,200 hours of driving.

To date, many studies have been conducted with the Waymo Open Dataset and the Lyft Level-5 dataset. For example, Wen et al. (2023) explored the strategies human drivers took when following AVs versus human-driven vehicles based on the trajectory data from the Waymo dataset. Similarly, T. Li et al. (2023) extracted human-following AV and human-following human events in the Lyft Level-5 dataset. However, it should be noted that the original objectives of the three datasets were for AV control algorithm development and were not designed for mixed-traffic behavioral research. For certain reasons, all these datasets chose to cut the video clips into 20-second-long discrete segments, and the information other than the location of the surrounding road agents has been removed. As a result, these datasets are not suitable for long-duration driver behavior extraction or complex scenarios that may exceed 20 seconds. Similarly, due to the sparsity of the crashes on the road, the trajectory-oriented datasets are not suitable for research that aims to understand the impact of AVs on traffic safety and analyze the patterns of AV-involved crashes. Thus, local authorities and governments have also initiated plans to collect information regarding AV-involved crashes. The two most widely used datasets include the NHTSA dataset (by the National Highway Traffic Safety Administration) and the CA-DMV dataset (by the California Department of Motor Vehicle), both involved data regarding

AV-involved crashes. Detailed information on these two datasets and other available datasets are listed in Table 2.3.

However, although the data from the observational studies can maintain the highest level of validity, it is not omnipotent. Specifically, given that the researchers usually have no access to the road users involved in certain traffic events, it is usually unlikely to collect information regarding the road users' psychological states or processes. Hence, the data from this type of research can hardly support the analysis of psychological or cognitive factors of human behaviors (e.g., how users may allocate their attention differently when interacting with AVs). Further, as the data was reported by the AV fleet operators, drivers, or police, the quality of the data might be compromised, and minor events (without leading to large property or injuries) may be under-reported.

2.2.5 Choice of Approaches for Mixed Traffic Research

In general, no single research method so far can support research on all types of research questions in mixed traffic. It is vital to understand the pros and cons of the research approaches when they are used for specific research questions. In addition to the monetary and time costs and constraints of the available resources, we may select the research approaches by considering the factors in the following three dimensions.

First, the validity of the data for specific research questions. For example, given that the data from the driving simulators have lower ecological validity compared to the data from the on-road studies, the data collected from on-road studies may be more suitable for developing quantitative models to inform the design of the AV algorithms.

Second, the availability of the data. Not all approaches can provide all the necessary information for all research questions. For example, although observational studies may provide more realistic data, the collection of demographic information and psychological states of road users would be difficult. While naturalistic driving studies can capture both realistic driving behaviors and demographic information of the drivers, the human-AV interaction samples might be too sparse at this stage, given that only a few regions or states have allowed the AVs to be operated on public roads. Hence, simulator studies or field experiments may better reveal how psychological factors may be associated with drivers' behaviors in mixed traffic.

Third, the controllability of the experimental conditions. For example, in both naturalistic and observational studies, experimenters can hardly control the traffic scenarios. Hence, large datasets are needed to extract rare events that may meet their criteria. In this case, the simulator or field study approach can be a better option to obtain data in a relatively short period of time, given that a higher level of experiment control can be guaranteed. Experimental control is also vital to identify the influential factors of certain behaviors, as consistent scenarios can reduce the variations uncounted by the factors of interest.

Regarding the research approaches that can be used for mixed traffic research, we can summarize their pros and cons in these three dimensions in Table 2.4. It should be noted that the pros and cons of the research methods may vary with the progress

TABLE 2.3
Available public crash/trajectory datasets involving driving automation

Datasets	Data collection duration	Types of involved road agents	Locations of data collection	Total length of data	Type of data	Related research
CA DMV	2014–2019	Conventional and high-level autonomous vehicles	California	712 autonomous vehicle collision reports received	AV collision and disengagement records	(Sinha et al., 2021)
NHTSA AV crash data archive	2008–2009	Conventional and high-level autonomous vehicles	Several cities in US	Over 1,100 autonomous vehicle crash records	Accidents records	(National Highway Traffic Safety Administration, 2013)
AV crash dataset	2014–2024	Conventional and autonomous vehicles (with ADAS)	Several cities in United States	Total of 2,236 pieces of data	Crash data from CA DMV, NHTSA, news and land use, weather, and geometry information	(Transport research center, 2024)
NGSIM	2016	Conventional vehicle	Emeryville, Los Angeles, and Atlanta	2 hours 30 minutes	Video data	(U.S. Department of Transportation Federal Highway Administration, n.d.)
HighD	2018	Conventional vehicle	German Highways	147 hours	Video data	(Krajewski et al., 2018)
Lyft Level 5	2019	Autonomous vehicle	Palo Alto, California	1,000 hours	Video and radar data	(Kesten et al., n.d.)
Waymo	2021–2024	Autonomous vehicle	25 Cities in US	10 hours 50 minutes	Video and radar data	(K. Chen et al., 2024)
nuScenes	2018–2023	Autonomous vehicle	Boston and Singapore	5 hours 33 minutes	Video & radar data	(motional, 2023)

TABLE 2.4
Comparisons of the research approaches for mixed traffic research

Approaches	Data validity	Data availability	Experimental control
Driving simulation	Low to medium	Depends on the needs of the research questions	High
Field study	Medium		Medium to high
On-road study	Medium to high		Low to medium
Naturalistic driving research	High		Low
Observational approach	High		Low

of time. For example, when more AVs are operating on public roads, it is possible to collect enough data with naturalistic driving studies.

2.3 DATA TYPE AND DATA PROCESSING IN EMPIRICAL RESEARCH

One of the basic questions one may ask when selecting the approaches is the types of data that can be collected with the approach. In general, for mixed traffic research, the following types of data were used in previous research (i.e., trajectory data, behavioral data, subjective data, and eye-tracking data).

2.3.1 Trajectory Data

2.3.1.1 Application of the Trajectory Data

The trajectory data can capture the dynamic processes of interactions among road agents in mixed traffic. In general, previous research used the trajectory data in two ways. The raw and cleaned trajectory data was more commonly used for research that focused on AV algorithm designs (e.g., motion planning). For example, Gu et al. (2020) used Waymo trajectory data to design a car following an algorithm based on a modified Long-Short-Term-Memory (LSTM). At the same time, some other studies relied more on the metrics extracted from the trajectory data. For example, in Wen et al. (2023), the headway distance and vehicle speed extracted from the trajectories of the leading and following vehicles were used to model the car-following behaviors of human drivers in mixed traffic. In another study, again using the Waymo dataset, T. Li et al. (2023) analyzed and compared how drivers responded to human-driven vehicles versus AVs differently by modeling the effect of road agent types (AV or human-driven vehicles) on the drivers' selection of headways in car-following and overtaking events. The trajectory data can also be used to evaluate the safety of mixed traffic. Given the sparsity of the AV-involved crashes, previous research also extracted surrogate safety metrics from the trajectory data. For example, Q.-L. Lu et al. (2021) compared the performance of different algorithms for collision prediction in mixed traffic based on the estimated TTC in the lane-changing event.

2.3.1.2 Data Processing for Trajectory Data

As mentioned previously, the trajectory data was mostly collected on the road (either public or closed-track), for example, from field studies (Huang et al., 2020), on-road studies (Y. Chen et al., 2020), naturalistic studies (G. Li et al., 2023), and observational studies (Basu & Saha, 2022). Thus, the data may contain lots of noise and, if used inappropriately, may lead to inaccurate model training or behavioral predictions in follow-up studies (M. Zhou et al., 2017). Hence, the raw trajectory data needs to be processed before being used. The main purpose of data processing for trajectory data is as follows:

- **Data format conversion**: Data collected by different sensors may have different data formats, so they need to be converted into a unified format for subsequent data processing and analysis. Special techniques may be required to obtain the necessary data. For example, computer visions (e.g., image resizing (Khajeh Hosseini et al., 2022), color space conversion (Baskaran et al., 2017), and feature extraction (Samanta et al., 2018)) may be required to extract road agents and identify their locations and movements from the video and image data. Similarly, pre-processing is also needed for the LiDAR data before trajectories can be extracted (Llorca et al., 2010; Yan et al., 2011).
- **Data annotation**: The data collected by some sensors may need to be annotated for training and testing purposes. For example, for image or video data, the objects and road signs may need to be annotated manually or semi-automatically before the trajectory can be extracted; for LiDAR data, it is also necessary to annotate the position and size of the objects. Yan et al. (2011) explore how to semantically annotate different types of trajectory data, this includes annotating image, video, or LiDAR data for training and testing purposes. It introduces a framework or methodology to semantically annotate these heterogeneous trajectory data to provide a basis for subsequent data processing and analysis. Llorca et al. (2010) deals with methods for the collection and processing of traffic data in V2I networks. This includes the annotation of sensor data to facilitate the enhancement and analysis of floating vehicle data. C. Chen et al. (2020) proposed a framework for representing and compressing vehicle trajectory data, in which the data collected by the sensors will be labeled to facilitate the efficient representation and compression of vehicle trajectory data.
- **Outlier detection and handling**: Abnormal situations (e.g., sensor failure or extreme environmental conditions) may lead to errors or strong noise in the collected data. In such cases, outlier detection and processing of data may be required to eliminate the strong influence of the outliers. Typical outlier detection methods include:
  - Z-score method (Hodge & Austin, 2004): This method identifies outliers by calculating the z-score of each data point and flagging those that fall outside a certain threshold.
  - Interquartile range (IQR) method (Yang et al., 2019): This method involves calculating the IQR of the data and identifying outliers as those that fall

below the first quartile minus 1.5 times the IQR or above the third quartile plus 1.5 times the IQR.
  - Modified z-score method (Sandbhor & Chaphalkar, 2019): Similar to the z-score method, it is more robust to outliers and works better with skewed distributions.
  - Mahalanobis distance (Ghorbani, 2019): This method calculates the distance of each data point from the centroid of the data distribution and flags those that fall beyond a certain threshold.
  - Local outlier factor (LOF) (Breunig et al., 2000): This method compares the density of data points in their local neighborhood to identify outliers.
- **Data imputation**: In data acquisition and transmission processes, some sensor data may be missing or removed due to errors. Thus, interpolation and other methods may be used to impute missing values. Typical interpolation methods include:
  - Linear interpolation (Kuffel et al., 1997): This method estimates missing values by drawing a straight line between the two nearest known data points and using the position along that line to estimate the missing value.
  - Polynomial interpolation (de Boor & Ron, 1990): This method fits a polynomial function to the known data points and uses it to estimate missing values.
  - Spline interpolation (De Boor, 1968): This method uses a piecewise polynomial function to interpolate between data points, providing a smooth estimate for missing values.
  - Kriging (van Beers & Kleijnen, 2004): This method is commonly used in geostatistics and spatial analysis to estimate unknown values based on the spatial correlation between known data points.
  - Moving average interpolation (Thompson, 1947): This method replaces missing values with the average of neighboring data points, which can help smooth out data fluctuations.

2.3.2 Behavioral Data

2.3.2.1 Application of the Behavioral Data

Although the trajectory data can reveal the interactions among the road agents, it may only provide superficial information regarding the outcome of road agents' decisions, but not how and why human road agents adopted the strategy. As a direct measure of how human road agents respond to AVs or other human-operated road agents on the road, the behavioral data may provide more insights into the impact of AVs on traffic safety and efficiency. For example, when investigating how pedestrians may interact with AVs, in addition to deep learning approaches that focused on trajectory prediction (Bhujel & Yau, 2023), researchers also focused on the pedestrians' hesitation time and body movements when encountering an AV (Rodríguez Palmeiro et al., 2018). In combination with questionnaire data, theoretical models regarding the relationships between the psychological states and demographic features of road users and their behaviors could be established. For example, Papadimitriou et al. (2016) delve into the impact of various psychological variables on pedestrian decision-making and interactions with traffic signals. This study can provide valuable insights into understanding how psychological factors influence pedestrians' behavior in urban environments, which can guide future endeavors to guide road users in mixed traffic.

2.3.2.2 Data Processing for Behavioral Data

Similar to the trajectory data, the behavioral data can be highly noisy. The first step in extracting behavioral data is the clear definition of the specific behaviors. Usually, this includes the pre-conditions for an action (e.g., only after the AV reaches 50 meters, a leg movement can be considered as a street crossing intention (Pillai, 2017)), the definition of the action (e.g., the movement of any leg or the forward-leaning of the upper body), and the time frame that an action can be counted (e.g., only the actions happened between the first glance to the cue and the event onset can be regarded as a pre-event action (He et al., 2021)).

Though we would try to clearly define the behaviors of interest, subjectivity inevitably has to be involved in the definition of certain behaviors. For example, in He et al. (2021), pre-event actions can be defined as any movement that is deemed as preparation for an upcoming event. Though the pre-conditions and the time frame can be strictly defined, the actions are subject to raters' judgment. Thus, to reduce the subjectivity in the behavior extraction, multiple raters would be involved. Preliminary extraction would be conducted to refine the criteria for behaviors, followed by discussions to reach a consensus and then independent judges by each rater.

However, given that the judgment of the behavioral data from each rater can be subjective, a common standard for evaluating the quantification of the behavior identification is inter-rater reliability (IRB). Typical IRB metrics include Cohen's Kappa (Ihejirika et al., 2015) and the intra-class correlation coefficient (Bobak et al., 2018). If the IRB metrics are lower than acceptable, then a refinement of the criteria is needed, and a re-extraction of behaviors based on the new standard may need to be conducted unless the metrics reach the threshold.

2.3.3 Subjective Data

2.3.3.1 Application of the Subjective Data

With behavioral data, although we can look into one's decisions in specific scenarios, it is still difficult to answer why the behavior differs and how we can shape one's behaviors. Further, the behavioral data can still hardly inform one's cognitive procedures leading to actions. Thus, to better model road users' behaviors, subjective data can be collected. For example, with the trajectory data only, we can rebuild their behavioral patterns, but we can hardly answer why human drivers took different strategies when following AVs versus following human-driven vehicles. Thus, it is difficult for us to design countermeasures to guide road users' behaviors in mixed traffic. To resolve this issue, a previous study (Zhao et al., 2020) rebuilt car-following scenarios in closed tracks; in addition, they also evaluated users' trust in AVs using a subjective questionnaire. With such an approach, the relationship between a psychological state (i.e., trust) and the car-following behaviors can be established.

However, so far, a gap between the psychological findings and the data-driven approach still exists. More specifically, although we can identify how subjective metrics can affect road users' behaviors or even build theoretical models explaining the variance accounted for by some psychological factors, these qualitative findings are not well utilized in the data-driven approach and can hardly guide the design of the AV control algorithms. Future research is still needed to bridge the gap between the psychology-based approach and the trajectory-/behavior-related approach.

2.3.3.2 Data Processing for Subjective Data

In most cases, the subjective data can be collected through questionnaires, interviews, and focus groups. The subjective data can be collected before, during, and after an experiment. Given that subjective data is related to what the participants feel or subjectively perceive the scenarios, the validity and reliability of the data might be questioned and need to be evaluated, especially if a self-designed questionnaire is used. Here, the validity reflects whether the subjective measure can accurately measure what it is supposed to measure, which includes three dimensions (i.e., construct validity, content validity, and criterion validity), while the reliability evaluates to what extent the measure can collect the same data in repeated observations, which also has three dimensions (i.e., test–retest reliability, interrater reliability, and internal reliability). Some typical metrics regarding validity and reliability include Cronbach's α (Tavakol & Dennick, 2011) and KMO (Prato et al., 2005). The readers could refer to the reference for more information regarding evaluating the validity and reliability. For the design of the questionnaire, interview, or focus group, the readers could refer to (Gubrium & Holstein, 2002; Stanton et al., 2017) for more information.

2.3.4 Physiological Data and Eye-Tracking Data

2.3.4.1 Application of the Physiological Data and Eye-Tracking Data

As the subjective data is sensitive to users' bias and the validity can sometimes be questionable, more objective data that can reflect road users' psychological states is sometimes preferred. It is widely acknowledged that the changes in some psychological states can lead to variations in physiological indices. For example, stress is associated with variations in the galvanic skin responses (Fernandes et al., 2014), heart rate, and heart rate variability (Kim et al., 2018; Taelman et al., 2009). Hence, previous research also adopted physiological measures to evaluate users' acceptance of external human–machine interface (eHMI) in mixed traffic (e.g., Tan et al., 2022). The electroencephalogram (EEG) has also been widely used in previous research. By monitoring the electrical activity of the brain, researchers can infer a driver's or pedestrian's emotional state and cognitive load when they are faced with stress, emergencies, or changes in comfort, which can further be used to infer road users' capability to respond to traffic incidents (P. Li et al., 2022; Luque et al., 2024). Physiological signals can also be combined with other sensor data for comprehensive analysis (Rao et al., 2023).

As for the eye-tracking measures, it is estimated that around 90% of needed information in driving is visual information (Sivak, 1998). Thus, understanding road users' road behaviors can provide insights into their attention allocation strategies. Eye trackers are widely adopted to collect road users' visual behaviors. With eye trackers, experimenters can identify the gaze movements of human road agents and reveal their attention allocation and perception patterns, thus enabling a deep understanding of their behaviors. For example, eye-tracking technologies have been widely used in pedestrian–AV interaction studies to understand how the eHMI design can influence the pedestrian's perception of the AVs and, hence, behaviors when encountering AVs. In addition, the shared attention between human road users and AVs may facilitate more efficient and safe interactions among them, given that the computer vision domain has already emphasized the importance of directing attention to areas of importance in traffic scenarios (e.g., Makrigiorgos et al., 2019).

2.3.4.2 Data Processing for Physiological Data and Eye-Tracking Data

The physiological data is subjective to noises in the environment. Thus, filtering noise is usually required as the first step in processing physiological data. For example, the electrocardiogram (ECG) may have some baseline drift (see Figure 2.2), which would downgrade the detection of the R peaks in the signal. Thus, as common practice, the raw signal would be detrended before being fed into the following steps. The EEG signal is also sensitive to the artifacts and the background noise in the environment. Thus, in addition to collecting data in the magnetic shielding room, the bandpass filter is usually used to filter out the utility frequency (e.g., 50 Hz in China and 60 Hz in the U.S.) in the EEG signals. After pre-processing, depending on the way the signals are used, specific metrics can be extracted from the raw signal, or the cleaned signals can be fed into some algorithms directly. For example, the inter-beat-intervals (iBi) can be extracted from the ECG signal and be further processed into the heart rate and the heart rate variability (HRV). The HRV can then be transformed into the frequency domain, and metrics such as the LF (power of the low-frequency) and HF (power of the high frequency) can be extracted. Alternatively, the raw ECG data can be fed into the algorithms directly

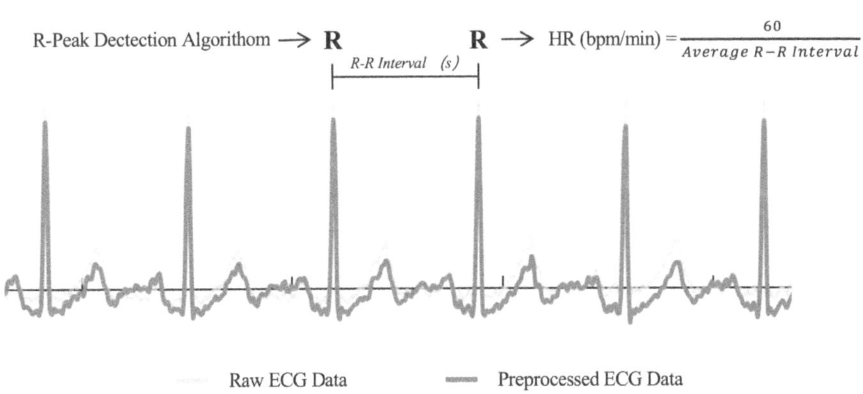

FIGURE 2.2 Typical ECG data.

without extracting handcrafted metrics; this is especially common in the deep learning domain so that the models might be able to learn the high-level features in the raw data (Xu et al., 2019).

The eye-tracking data includes two categories: gaze-related data and non-gaze-related data. As can be inferred from the name, the gaze-related data is about the location where the road agents are looking. This category of data is directly related to users' attention allocation. A number of metrics can be extracted from the gaze-related data, including but not limited to glance duration, fixation duration, frequency of the glances, percentage of time looking at a specific area of interest, and spatial density. For more information regarding the definition of gaze-related data in the driving domain, the readers could refer to the ISO standards (e.g., ISO_PRF_15007-1_(E) and ISO_TS_15007-2_(E)). At the same time, the non-gaze-related data also contains rich information regarding users' states. For example, the percentage time of eye closure (PERCLOS) is closely related to one's fatigue levels (J. Zhang et al., 2021), and the size of the pupil is sensitive to the workload one is experiencing (Klingner et al., 2008).

2.4 EXPERIMENT DESIGN BASICS FOR EMPIRICAL RESEARCH

After deciding the types of data to be collected, another key task in conducting empirical studies is the experiment design. In general, when designing an empirical study, the experimenters need to consider three aspects of techniques, that is, how to **measure** the variables of interest, how to **manipulate** some property of an actor-behavior-context (i.e., making a variable have a particular predetermined value or level for certain scenarios to be studied, and other specific values or levels for certain other scenarios so that the effect of the variable can be assessed by comparing those two sets of scenarios), and how to **control** the impact of variables that are important but irrelevant in a particular study.

In general, the experiment design can be regarded as balancing the pros and cons of different techniques. For example, to understand the influence of two eHMI designs on pedestrians' street crossing behavior, we can either let one participant experience multiple eHMIs (a within-subject design) or have multiple participants experiencing different eHMIs respectively (a between-subjects design). If a between-subjects design is adopted, then we may need to worry if there are individual differences between participants in different groups that may shadow the effect of eHMI so that the matching technique will need to be considered (i.e., making sure specific features of the participants across groups to be similar). However, in some scenarios, we may not know which features would influence their behaviors, and thus, it is difficult to match the participants across groups. Alternatively, if a within-subject design is adopted, we may need to worry about the learning effect and the fatigue effect (i.e., participants might learn to perform better or become tired with the progress of the experiment) if multiple trials were conducted for each participant. Thus, the counterbalancing technique will need to be used. However, this will increase the number of required participants, as a full counterbalance would significantly increase the required number of participants (e.g., a minimum of 6 participants for three levels,

24 orders for four levels, and 120 orders for five levels). In summary, the experiment design is more than an art rather than a technique, and that is also why multiple empirical studies may be needed for researchers to draw relatively solid conclusions on a research question. For more information regarding the experiment design and the following data analysis, especially the choice of the statistical models, the readers could refer to other books (e.g., Experimental Design: Procedures for the Behavioral Sciences by Maydeu-Olivares & Millsap (2009)).

REFERENCES

Ariansyah, D., Pardamean, B., & Caruso, G. (2023). The effect of visual advanced driver assistance systems on a following human driver in a mixed-traffic condition. *Procedia Computer Science*, 216, 221–229. https://doi.org/10.1016/j.procs.2022.12.130

Baskaran, S., Fang, S., & Jiang, S. (2017). Spatiotemporal visualization of traffic paths using color space time curve. In *2017 IEEE International Conference on Big Data (Big Data)*, 3398–3405. https://doi.org/10.1109/BigData.2017.8258325

Basu, S., & Saha, P. (2022). Evaluation of risk factors for road accidents under mixed traffic: Case study on Indian highways. *IATSS Research*, 46(4), 559–573. https://doi.org/10.1016/j.iatssr.2022.09.004

Bhujel, N., & Yau, W.-Y. (2023). Disentangling crowd interactions for pedestrians trajectory prediction. *IEEE Robotics and Automation Letters*, 8(5), 3078–3085. https://doi.org/10.1109/LRA.2023.3262420

Bobak, C. A., Barr, P. J., & O'Malley, A. J. (2018). Estimation of an inter-rater intra-class correlation coefficient that overcomes common assumption violations in the assessment of health measurement scales. *BMC Medical Research Methodology*, 18(1), 93. https://doi.org/10.1186/s12874-018-0550-6

Breunig, M. M., Kriegel, H.-P., Ng, R. T., & Sander, J. (2000). LOF: Identifying density-based local outliers. In *Proceedings of the 2000 ACM SIGMOD International Conference on Management of Data*, 93–104. https://doi.org/10.1145/342009.335388

Chen, C., Ding, Y., Guo, S., & Wang, Y. (2020). DAVT: An error-bounded vehicle trajectory data representation and compression framework. *IEEE Transactions on Vehicular Technology*, 69(10), 10606–10618. https://doi.org/10.1109/TVT.2020.3015214

Chen, K., Ge, R., Qiu, H., Al-Rfou, R., Qi, C. R., Zhou, X., Yang, Z., Ettinger, S., Sun, P., Leng, Z., Baniodeh, M., Bogun, I., Wang, W., Tan, M., & Anguelov, D. (2024). WOMD-LiDAR: Raw sensor dataset benchmark for motion forecasting (arXiv:2304.03834). *arXiv*. https://doi.org/10.48550/arXiv.2304.03834

Chen, Y., Chen, S., Xiao, T., Zhang, S., Hou, Q., & Zheng, N. (2020). Mixed test environment-based vehicle-in-the-loop validation—a new testing approach for autonomous vehicles. In *2020 IEEE Intelligent Vehicles Symposium (IV)*, 1283–1289. https://doi.org/10.1109/IV47402.2020.9304658

Chityala, S., Sobanjo, J. O., Erman Ozguven, E., Sando, T., & Twumasi-Boakye, R. (2020). Driver behavior at a freeway merge to mixed traffic of conventional and connected autonomous vehicles. *Transportation Research Record: Journal of the Transportation Research Board*, 2674(11). https://trid.trb.org/View/1739354

Dahlbäck, N., Jönsson, A., & Ahrenberg, L. (1993). Wizard of Oz studies—why and how. *Knowledge-Based Systems*, 6(4), 258–266. https://doi.org/10.1016/0950-7051(93)90017-N

de Boor, C. (1968). On the convergence of odd-degree spline interpolation. *Journal of Approximation Theory*, 1(4), 452–463. https://doi.org/10.1016/0021-9045(68)90033-6

de Boor, C., & Ron, A. (1990). On multivariate polynomial interpolation. *Constructive Approximation*, 6(3), 287–302. https://doi.org/10.1007/BF01890412

Detjen, H., Pfleging, B., & Schneegass, S. (2020). A Wizard of Oz field study to understand non-driving-related activities, trust, and acceptance of automated vehicles. In *12th International Conference on Automotive User Interfaces and Interactive Vehicular Applications*, 19–29. https://doi.org/10.1145/3409120.3410662

Eenink, R., Barnard, Y., Baumann, M., Augros, X., & Utesch, F. (2014, April). UDRIVE: The European naturalistic driving study [Proceedings Paper]. TRA 2014. In *Proceedings of Transport Research Arena; IFSTTAR*. http://tra2014.traconference.eu/papers/pdfs/TRA2014_Fpaper_19523.pdf

Fernandes, A., Helawar, R., Lokesh, R., Tari, T., & Shahapurkar, A. V. (2014). Determination of stress using blood pressure and galvanic skin response. In *2014 International Conference on Communication and Network Technologies*, 165–168. https://doi.org/10.1109/CNT.2014.7062747

Fuest, T., Feierle, A., Schmidt, E., & Bengler, K. (2020). Effects of marking automated vehicles on human drivers on highways. *Information*, 11(6), 286. https://doi.org/10.3390/info11060286

Ghorbani, H. (2019). Mahalanobis distance and its application for detecting multivariate outliers. *Facta Universitatis, Series: Mathematics and Informatics*, 34(3), 583–595. https://doi.org/10.22190/FUMI1903583G

Gu, Z., Li, Z., Di, X., & Shi, R. (2020). An LSTM-based autonomous driving model using a Waymo Open Dataset. *Applied Sciences*, 10(6), 2046. https://doi.org/10.3390/app10062046

Gubrium, J. F., & Holstein, J. A. (2002). *Handbook of Interview Research: Context and Method*. SAGE.

Guo, F., Han, S., & Hankey, J. (2022). *The Shanghai Naturalistic Driving Study (SHNDS)*.

Hallmark, S. L., Tyner, S., Oneyear, N., Carney, C., & McGehee, D. (2015). Evaluation of driving behavior on rural 2-lane curves using the SHRP 2 naturalistic driving study data. *Journal of Safety Research*, 54, 17–27. https://doi.org/10.1016/j.jsr.2015.06.017

He, D., Kanaan, D., & Donmez, B. (2021). In-vehicle displays to support driver anticipation of traffic conflicts in automated vehicles. *Accident Analysis & Prevention*, 149, 105842. https://doi.org/10.1016/j.aap.2020.105842

Hodge, V., & Austin, J. (2004). A survey of outlier detection methodologies. *Artificial Intelligence Review*, 22(2), 85–126. https://doi.org/10.1023/B:AIRE.0000045502.10941.a9

Hong, Y., Min, J., King, C. B., & Meeker, W. Q. (2021). Reliability analysis of artificial intelligence systems using recurrent events data from autonomous vehicles. *arXiv (arXiv:2102.01740)*. http://arxiv.org/abs/2102.01740

Hua, Q., Jin, L., Jiang, Y., Guo, B., & Xie, X. (2021). Effect of cognitive distraction on physiological measures and driving performance in traditional and mixed traffic environments. *Journal of Advanced Transportation*, 2021, 1–17. https://doi.org/10.1155/2021/6739071

Huang, Y., Ding, H., Zhang, Y., Wang, H., Cao, D., Xu, N., & Hu, C. (2020). A motion planning and tracking framework for autonomous vehicles based on artificial potential field elaborated resistance network approach. *IEEE Transactions on Industrial Electronics*, 67(2), 1376–1386. https://doi.org/10.1109/TIE.2019.2898599

Ihejirika, R. C., Thakore, R. V., Sathiyakumar, V., Ehrenfeld, J. M., Obremskey, W. T., & Sethi, M. K. (2015). An assessment of the inter-rater reliability of the ASA physical status score in the orthopaedic trauma population. *Injury*, 46(4), 542–546. https://doi.org/10.1016/j.injury.2014.02.039

Janatabadi, F., & Ermagun, A. (2022). Empirical evidence of bias in public acceptance of autonomous vehicles. *Transportation Research Part F: Traffic Psychology and Behaviour*, 84, 330–347. https://doi.org/10.1016/j.trf.2021.12.005

Jiang, L., Ji, J., Ren, Y., Wang, H., & Huang, Y. (2020). Risk modeling and quantification of a platoon in mixed traffic based on the mass-spring-Damper Model. *Journal of Advanced Transportation, 2020*, Article ID 7475682. https://trid.trb.org/View/1722395

Kalantari, A. H., Yang, Y., Garcia de Pedro, J., Lee, Y. M., Horrobin, A., Solernou, A., Holmes, C., Merat, N., & Markkula, G. (2023). Who goes first? A distributed simulator study of vehicle-pedestrian interaction. *Accident; Analysis and Prevention*, 186, 107050. https://doi.org/10.1016/j.aap.2023.107050

Kesten, R., Usman, M., Houston, J., & Pandya, T. (n.d.). Lyft level 5 AV dataset 2019. Retrieved June 15, 2024, from http://archive.org/details/lyftlevel5avdataset2019

Khajeh Hosseini, M., Talebpour, A., Devunuri, S., & Hamdar, S. H. (2022). An unsupervised learning framework for detecting adaptive cruise control operated vehicles in a vehicle trajectory data. *Expert Systems with Applications*, 208, 118060. https://doi.org/10.1016/j.eswa.2022.118060

Kim, H.-G., Cheon, E.-J., Bai, D.-S., Lee, Y. H., & Koo, B.-H. (2018). Stress and heart rate variability: A meta-analysis and review of the literature. *Psychiatry Investigation*, 15(3), 235–245. https://doi.org/10.30773/pi.2017.08.17

Klauer, C., Pearson, J., & Hankey, J. (2018). An overview of the Canada naturalistic driving and Canada truck naturalistic driving studies. In *7th International Symposium on Naturalistic Driving Research Inn at Virginia Tech*, Blacksburg, VA.

Klingner, J., Kumar, R., & Hanrahan, P. (2008). Measuring the task-evoked pupillary response with a remote eye tracker. In *Proceedings of the 2008 Symposium on Eye Tracking Research & Applications*, 69–72. https://doi.org/10.1145/1344471.1344489

Krajewski, R., Bock, J., Kloeker, L., & Eckstein, L. (2018). The highD dataset: A drone dataset of naturalistic vehicle trajectories on German highways for validation of highly automated driving systems. In *2018 21st International Conference on Intelligent Transportation Systems (ITSC)*, 2118–2125. https://doi.org/10.1109/ITSC.2018.8569552

Kuffel, P., Kent, K., & Irwin, G. (1997). The implementation and effectiveness of linear interpolation within digital simulation. *International Journal of Electrical Power & Energy Systems*, 19(4), 221–227. https://doi.org/10.1016/S0142-0615(96)00045-2

Lee, S. E., Simons-Morton, B. G., Klauer, S. E., Ouimet, M. C., & Dingus, T. A. (2011). Naturalistic assessment of novice teenage crash experience. *Accident Analysis & Prevention*, 43(4), 1472–1479. https://doi.org/10.1016/j.aap.2011.02.026

Li, G., Jiao, Y., Knoop, V. L., Calvert, S. C., & Van Lint, J. W. C. (2023). Large Car-following Data Based on Lyft level-5 Open Dataset: Following autonomous vehicles vs. human-driven vehicles. In *2023 IEEE 26th International Conference on Intelligent Transportation Systems (ITSC)*, 5818–5823. https://doi.org/10.1109/ITSC57777.2023.10422574

Li, P., Li, Y., Yao, Y., Wu, C., Nie, B., & Li, S. E. (2022). Sensitivity of electrodermal activity features for driver arousal measurement in cognitive load: The application in automated driving systems. *IEEE Transactions on Intelligent Transportation Systems*, 23(9), 14954–14967. https://doi.org/10.1109/TITS.2021.3135266

Li, T., Han, X., Ma, J., Ramos, M., & Lee, C. (2023). Operational safety of automated and human driving in mixed traffic environments: A perspective of car-following behavior. *Proceedings of the Institution of Mechanical Engineers, Part O: Journal of Risk and Reliability*, 237(2), 355–366. https://doi.org/10.1177/1748006X211050696

Liu, Y., Abediasl, H., Ansari, A., & Shahbakhti, M. (2023). Characterizing driving behavior and link to fuel consumption for university campus shuttle minibuses. Canadian Society for Mechanical Engineering (CSME) / Computational Fluid Dynamics Society of Canada

(CFD Canada). *International Congress (2023: Université de Sherbrooke, Québec)*, 6, ID 212. https://savoirs.usherbrooke.ca/handle/11143/21171

Llorca, D. F., Sotelo, M. A., Sánchez, S., Ocaña, M., Rodríguez-Ascariz, J. M., & García-Garrido, M. A. (2010). Traffic data collection for floating car data enhancement in V2I networks. *EURASIP Journal on Advances in Signal Processing*, 2010(1), 719294. https://doi.org/10.1155/2010/719294

Lu, J., Hossain, S., Sheng, W., & Bai, H. (2023). Cooperative driving in mixed traffic of manned and unmanned vehicles based on human driving behavior understanding. In *2023 IEEE International Conference on Robotics and Automation (ICRA)*, 3532–3538. https://doi.org/10.1109/ICRA48891.2023.10160282

Lu, J., Stracener, R., Sheng, W., Bai, H., & Hossain, S. (2022). Development of a research testbed for cooperative driving in mixed traffic of human-driven and autonomous vehicles. In *2022 IEEE/RSJ International Conference on Intelligent Robots and Systems (IROS)*, 11403–11408. https://doi.org/10.1109/IROS47612.2022.9981436

Lu, Q.-L., Yang, K., & Antoniou, C. (2021). Crash risk analysis for the mixed traffic flow with human-driven and connected and autonomous vehicles. In *2021 IEEE International Intelligent Transportation Systems Conference (ITSC)*, 1233–1238. https://doi.org/10.1109/ITSC48978.2021.9564897

Luque, F., Armada, V., Piovano, L., Jurado-Barba, R., & Santamaría, A. (2024). Understanding pedestrian cognition workload in traffic environments using virtual reality and electroencephalography. *Electronics*, 13(8), 1453. https://doi.org/10.3390/electronics13081453

Ma, Z., & Zhang, Y. (2024). Driver-automated vehicle interaction in mixed traffic: Types of interaction and drivers' driving styles. *Human Factors: The Journal of the Human Factors and Ergonomics Society*, 66(2), 544–561. https://doi.org/10.1177/00187208221088358

Makrigiorgos, A., Shafti, A., Harston, A., Gerard, J., & Faisal, A. A. (2019). Human visual attention prediction boosts learning & performance of autonomous driving agents. *arXiv (arXiv:1909.05003)*. https://doi.org/10.48550/arXiv.1909.05003

Maydeu-Olivares, A., & Millsap, R. E. (2009). *The SAGE Handbook of Quantitative Methods in Psychology* (pp. 1–800). R. E. Millsap (ed.). SAGE. https://doi.org/10.4135/9780857020994

Miller, L., Koniakowsky, I. M., Kraus, J., & Baumann, M. (2022). The impact of expectations about automated and manual vehicles on drivers' behavior: Insights from a mixed traffic driving simulator study. In *Proceedings of the 14th International Conference on Automotive User Interfaces and Interactive Vehicular Applications*, 150–161. https://doi.org/10.1145/3543174.3546837

Miller, L., Kraus, J., Koniakowsky, I., Pichen, J., & Baumann, M. (2023). Learning in mixed traffic: Drivers' adaptation to ambiguous communication depending on their expectations toward automated and manual vehicles. *International Journal of Human–Computer Interaction*, 39(16), 3268–3287. https://doi.org/10.1080/10447318.2023.2215097

motional. (2023). nuScenes. *Nuscenes*. www.nuscenes.org/

National Highway Traffic Safety Administration. (2013). Motor Vehicle Traffic Crashes as a Leading Cause of Death in the United States, 2008 and 2009. *Annals of Emergency Medicine*, 61(4), 484. https://doi.org/10.1016/j.annemergmed.2013.02.009

National Research Council (U.S.) (Ed.). (2001). *Strategic Highway Research: Saving Lives, Reducing Congestion, Improving Quality of Life*. Transportation Research Board, National Research Council, National Academy Press.

Olsson, N. (2023). A validation study comparing performance in a low-fidelity train-driving simulator with actual train driving performance. *Transportation Research Part F: Traffic Psychology and Behaviour*, 97, 109–122. https://doi.org/10.1016/j.trf.2023.07.007

Papadimitriou, E., Lassarre, S., & Yannis, G. (2016). Introducing human factors in pedestrian crossing behaviour models. *Transportation Research Part F: Traffic Psychology and Behaviour*, 36, 69–82. https://doi.org/10.1016/j.trf.2015.11.003

Pillai, A. (2017). Virtual reality based study to analyse pedestrian attitude towards autonomous vehicles. Master's thesis, Master's programme in ICT Innovation. https://aaltodoc.aalto.fi/handle/123456789/28563

Prato, C. G., Bekhor, S., & Pronello, C. (2005). Methodology for exploratory analysis of latent factors influencing drivers' behavior. *Transportation Research Record*, 1926(1), 115–125. https://doi.org/10.1177/0361198105192600114

Rao, S., Ghosh, S., Rodriguez, G. P., Röggla, T., Cesar, P., & El Ali, A. (2023). From video to hybrid simulator: Exploring affective responses toward non-verbal pedestrian crossing actions using camera and physiological sensors. *International Journal of Human–Computer Interaction*, 39(16), 3213–3236. https://doi.org/10.1080/10447318.2023.2224955

Rodríguez Palmeiro, A., van der Kint, S., Vissers, L., Farah, H., de Winter, J. C. F., & Hagenzieker, M. (2018). Interaction between pedestrians and automated vehicles: A Wizard of Oz experiment. *Transportation Research Part F: Traffic Psychology and Behaviour*, 58, 1005–1020. https://doi.org/10.1016/j.trf.2018.07.020

Samanta, S., Mukherjee, A., Ashour, A. S., Dey, N., Tavares, J. M. R. S., Abdessalem Karâa, W. B., Taiar, R., Azar, A. T., & Hassanien, A. E. (2018). Log transform based optimal image enhancement using firefly algorithm for autonomous mini unmanned aerial vehicle: An application of aerial photography. *International Journal of Image and Graphics*, 18(04), 1850019. https://doi.org/10.1142/S0219467818500195

Sandbhor, S., & Chaphalkar, N. B. (2019). Impact of outlier detection on neural networks based property value prediction. In S. C. Satapathy, V. Bhateja, R. Somanah, X.-S. Yang, & R. Senkerik (Eds.), *Information Systems Design and Intelligent Applications* (pp. 481–495). Springer. https://doi.org/10.1007/978-981-13-3329-3_45

Sinha, A., Chand, S., Vu, V., Chen, H., & Dixit, V. (2021). Crash and disengagement data of autonomous vehicles on public roads in California. *Scientific Data*, 8(1), 298. https://doi.org/10.1038/s41597-021-01083-7

Sivak, M. (1998). The information that drivers use: Is it indeed 90 percent visual? *The UMTRI Research Review*, 29(1), 1.

Stange, V., Kühn, M., & Vollrath, M. (2022a). Manual drivers' experience and driving behavior in repeated interactions with automated Level 3 vehicles in mixed traffic on the highway. *Transportation Research Part F: Traffic Psychology and Behaviour*, 87, 426–443. https://doi.org/10.1016/j.trf.2022.04.019

Stange, V., Kühn, M., & Vollrath, M. (2022b). Safety at first sight?—Manual drivers' experience and driving behavior at first contact with Level 3 vehicles in mixed traffic on the highway. *Transportation Research Part F: Traffic Psychology and Behaviour*, 87, 327–346. https://doi.org/10.1016/j.trf.2022.04.004

Stanton, N. A., Salmon, P. M., Rafferty, L. A., Walker, G. H., Baber, C., & Jenkins, D. P. (2017). *Human Factors Methods: A Practical Guide for Engineering and Design* (2nd ed.). CRC Press. https://doi.org/10.1201/9781315587394 .

Steve, S. (2011). *Intelligent Transportation Systems—Integrated Vehicle-Based Safety Systems (IVBSS)*. Washington, DC: Office of the Assistant Secretary for Research and Technology. https://its.dot.gov/research_archives/ivbss/index.htm

Taelman, J., Vandeput, S., Spaepen, A., & Van Huffel, S. (2009). Influence of mental stress on heart rate and heart rate variability. In J. Vander Sloten, P. Verdonck, M. Nyssen, & J. Haueisen (Eds.), *4th European Conference of the International Federation for Medical and Biological Engineering* (pp. 1366–1369). Springer. https://doi.org/10.1007/978-3-540-89208-3_324

Tan, Z., Dai, N., Su, Y., Zhang, R., Li, Y., Wu, D., & Li, S. (2022). Human–machine interaction in intelligent and connected vehicles: A review of status quo, issues, and opportunities. *IEEE Transactions on Intelligent Transportation Systems*, 23(9), 13954–13975. https://doi.org/10.1109/TITS.2021.3127217

Tavakol, M., & Dennick, R. (2011). Making sense of Cronbach's alpha. *International Journal of Medical Education*, 2, 53–55. https://doi.org/10.5116/ijme.4dfb.8dfd

Thompson, W. R. (1947). Use of moving averages and interpolation to estimate median-effective dose: I. Fundamental formulas, estimation of error, and relation to other methods. *Bacteriological Reviews*, 11(2), 115–145. https://doi.org/10.1128/br.11.2.115-145.1947

Transport Research Center. (2024). *AV Crashes | Autonomous Vehicle Crashes*. Autonomous Vehicle Crashes. www.avcrashes.net/

U.S. Department of Transportation Federal Highway Administration. (n.d.). Next generation simulation (NGSIM) open data. Next generation simulation (NGSIM) open data. Retrieved June 15, 2024, from https://datahub.transportation.gov/stories/s/Next-Generation-Simulation-NGSIM-Open-Data/i5zb-xe34/

van Beers, W. C. M., & Kleijnen, J. P. C. (2004). Kriging interpolation in simulation: A survey. In *Proceedings of the 2004 Winter Simulation Conference, 2004*, 1, 121. https://doi.org/10.1109/WSC.2004.1371308

Wen, X., Jian, S., & He, D. (2023). Modeling the effects of autonomous vehicles on human driver car-Following Behaviors Using Inverse Reinforcement Learning. *IEEE Transactions on Intelligent Transportation Systems*, 24(12), 13903–13915. https://doi.org/10.1109/TITS.2023.3298150

Williamson, A., Grzebieta, R., Eusebio, J. E., Zheng, W. Y., Wall, J., Charlton, J., Lenne, M., Haley, J., Barnes, B., Rakotonirainy, A., Woolley, J., Senserrick, T., Young, K., Haworth, N., Regan, M., Cockfield, S., Healy, D., Cavallo, A., Di Stefano, M., … Cornish, M. (2015). The Australian Naturalistic Driving Study: From beginnings to launch. In I. Cameron, N. Haworth, & L. McIntosh (Eds.), *Proceedings of the 2015 Australasian Road Safety Conference (ARSC2015)* (pp. 1–7). Australasian College of Road Safety (ACRS). http://acrs.org.au/files/papers/arsc/2015/WilliamsonA%20277%20The%20Australian%20naturalistic%20driver%20study.pdf

Wu, M., Wang, N., & Yuen, K. F. (2023). Can autonomy level and anthropomorphic characteristics affect public acceptance and trust towards shared autonomous vehicles? *Technological Forecasting and Social Change*, 189, 122384. https://doi.org/10.1016/j.techfore.2023.122384

Xu, S. S., Mak, M.-W., & Cheung, C.-C. (2019). Towards end-to-end ECG classification with raw signal extraction and deep neural networks. *IEEE Journal of Biomedical and Health Informatics*, 23(4), 1574–1584. https://doi.org/10.1109/JBHI.2018.2871510

Yan, Z., Chakraborty, D., Parent, C., Spaccapietra, S., & Aberer, K. (2011). SeMiTri: A framework for semantic annotation of heterogeneous trajectories. In *Proceedings of the 14th International Conference on Extending Database Technology* (pp. 259–270). Association for Computing Machinery. https://doi.org/10.1145/1951365.1951398

Yang, J., Rahardja, S., & Fränti, P. (2019). Outlier detection: How to threshold outlier scores? In *Proceedings of the International Conference on Artificial Intelligence, Information Processing and Cloud Computing* (pp. 1–6). https://doi.org/10.1145/3371425.3371427

Zhang, J., Chen, Z., Liu, W., Ding, P., & Wu, Q. (2021). A field study of work type influence on air traffic controllers' fatigue based on data-driven PERCLOS detection. *International Journal of Environmental Research and Public Health*, 18(22), 11937. https://doi.org/10.3390/ijerph182211937

Zhang, T., Tao, D., Qu, X., Zhang, X., Zeng, J., Zhu, H., & Zhu, H. (2020). Automated vehicle acceptance in China: Social influence and initial trust are key determinants. *Transportation Research Part C: Emerging Technologies*, 112, 220–233. https://doi.org/10.1016/j.trc.2020.01.027

Zhao, X., Wang, Z., Xu, Z., Wang, Y., Li, X., & Qu, X. (2020). Field experiments on longitudinal characteristics of human driver behavior following an autonomous vehicle. *Transportation Research Part C: Emerging Technologies*, 114, 205–224. https://doi.org/10.1016/j.trc.2020.02.018

Zhou, A., Liu, Y., Tenenboim, E., Agrawal, S., & Peeta, S. (2023). Car-following behavior of human-driven vehicles in mixed-flow traffic: A driving simulator study. *IEEE Transactions on Intelligent Vehicles*, 8(4), 2661–2673. https://doi.org/10.1109/TIV.2023.3257962

Zhou, M., Qu, X., & Li, X. (2017). A recurrent neural network based microscopic car following model to predict traffic oscillation. *Transportation Research Part C: Emerging Technologies*, 84, 245–264. https://doi.org/10.1016/j.trc.2017.08.027

3 Tools for Micro Mixed Traffic Research

Meixin Zhu

3.1 DATA-DRIVEN APPROACH

The current availability of high-fidelity traffic data and the resulting data-driven approach enables direct modeling of drivers' behavior using massive field data, exhibiting greater flexibility than traditional models. A data-driven approach, proposed to induce background vehicles in simulated environments to react to autonomous vehicle (AV) behaviors akin to human drivers on the road, offers distinct advantages over rule-based methods. These advantages include the automatic discovery of patterns within extensive datasets, resulting in a greater diversity of driving behaviors. Additionally, the data-driven approach excels in simulating more realistic driving behaviors, particularly in challenging scenarios like rear-end collisions, posing difficulties for rule-based expertise. Consequently, the majority of research has transitioned toward employing data-driven approaches, previously constrained by the limited availability of extensive datasets but alleviated by the emergence of numerous high-quality datasets. From a modeling perspective, data-driven approaches can be classified into four major types: imitation learning (IL) models, reinforcement learning (RL) models, deep generative models, and deep learning (DL) models.

3.1.1 Imitation Learning Models

IL plays a pivotal role in micro-mixed traffic studies, providing a valuable approach to modeling and simulating the behavior of diverse vehicles in complex traffic scenarios. In micro-mixed traffic, encompassing the interaction of various vehicle types such as cars, bicycles, and pedestrians, IL facilitates the development of intelligent agents capable of mimicking the behavior of human drivers and other road participants.

The utilization of IL in micro-mixed traffic studies encompasses training models to replicate the observed behavior of drivers in real-world scenarios. By leveraging expert demonstrations or data from actual traffic situations, these models learn to emulate the decision-making processes, maneuvers, and interactions of human drivers. This process enables the generation of realistic and adaptive simulations that capture the nuances of mixed traffic dynamics.

One notable advantage of IL in micro-mixed traffic studies is its ability to model diverse and complex driving behaviors. Unlike rule-based simulations that may face

 DOI: 10.1201/9781032674667-3

TABLE 3.1
A comparison of BC and GAIL

Methods	Advantages	Disadvantages
BC	*Deterministic Policy Learning:* BC typically learns a deterministic mapping from states to actions based on expert demonstrations. This can result in stable and predictable behavior.	*Lack of Adaptability to Novel Situations:* BC tends to perform poorly in situations that deviate significantly from the expert's demonstrations. It may struggle to handle unforeseen scenarios or situations not encountered during training.
	Reduced Exploration Effort: BC directly imitates the expert's actions, bypassing the need for costly exploration in the environment. This can lead to faster training times and more reliable policies.	*Limited Robustness to Noisy Data:* BC is sensitive to noise in expert demonstrations, which can lead to suboptimal policies. If the expert data contains errors or noise, BC may learn incorrect behaviors.
GAIL	*Improved Robustness to Noisy Data:* GAIL is more robust to noise in the expert demonstrations compared to BC. It can better filter out errors or noise in the expert data, leading to more accurate policy learning.	*Potential for Mode Collapse:* GAIL, like other GAN-based methods, can suffer from mode collapse, where the generator fails to capture the full diversity of expert behavior. This can lead to a limited range of generated traffic scenarios.

challenges in representing the variability in human actions, IL enables the automatic discovery of patterns within extensive datasets. This results in models capable of adapting to different driving styles, thereby enhancing the overall fidelity of micro-mixed traffic simulations.

Furthermore, IL contributes to tackling challenges in modeling the stochastic nature of human behavior on the road. It enables the simulation of more realistic driving behaviors, encompassing reactions to unexpected events and interactions between various road users. This is especially valuable in scenarios such as rear-end collisions, where the complexity of human behavior makes it challenging to capture using manually designed rules. There are two common ways to perform IL: behavior cloning (BC) and generative adversarial IL (GAIL). A comparison of them is shown in Table 3.1 (Chen, Zhu, Yang, Wang, & Wang, 2024).

3.1.2 Reinforcement Learning Models

The application of RL in micro-mixed traffic research entails harnessing this influential paradigm to cultivate intelligent agents capable of acquiring optimal decision-making policies in dynamic and intricate traffic environments. In the context of micro-mixed traffic, encompassing diverse road users such as cars, bicycles, and pedestrians, RL provides a principled approach to modeling adaptive and autonomous behavior. RL operates based on the principle of agents interacting with an environment and receiving feedback in the form of rewards or penalties according to their actions. RL agents learn to navigate the traffic environment through iterative decision-making, feedback reception, and strategy adjustment to maximize cumulative rewards.

RL agents, when trained on suitable datasets or in simulation environments, can learn to navigate through mixed traffic scenarios, taking into account factors like traffic flow, pedestrian movements, and interactions with other vehicles. RL also facilitates the investigation of traffic management strategies, as RL agents can undergo training to control traffic signals, manage intersections, and optimize traffic flow. This application becomes especially relevant in urban settings with intricate traffic dynamics, where RL-based traffic management systems can contribute to alleviating congestion and enhancing overall efficiency.

Furthermore, RL serves as a valuable tool for comprehending the emergence of collective behaviors in micro-mixed traffic scenarios. Through the interaction of RL agents, emergent traffic patterns and phenomena can be observed, contributing to an enhanced understanding of the intricate dynamics inherent in mixed traffic environments. Current research focuses on applying Deep Reinforcement Learning (DRL) methods and Adversarial Inverse Reinforcement Learning (AIRL) methods within traffic research. A comparison of the three methods when applied in traffic research is shown in Table 3.2 (Chen et al., 2024).

To ensure effective research, including realistic traffic models that align with human knowledge is crucial. However, incorporating human preferences for realism in traffic research remains challenging due to the limited expressiveness of current methods. The field of artificial intelligence (AI) has shown considerable interest in large language models (LLMs). Reinforcement learning with human feedback (RLHF) (Griffith, Subramanian, Scholz, Isbell, & Thomaz, 2013) has demonstrated potential for improving LLMs. Integrating LLMs into micro-mixed traffic research enables common sense reasoning (Cui et al., 2024), increasing the realism and complexity of generated traffic scenarios. In mixed traffic studies, finding the right balance between leveraging human expertise and maintaining the scalability and objectivity of RL algorithms is crucial. The combination of LLMs with RL has the potential to create more efficient, safe, and socially acceptable traffic systems, but careful implementation is required to avoid the pitfalls associated with LLM involvement.

3.1.3 Deep Generative Models

Deep generative models (DGMs) play a pivotal role in mixed traffic studies by providing a powerful framework for creating realistic and diverse scenarios. In the context of mixed traffic, which involves a multitude of road users such as cars, bicycles, and pedestrians, generative modeling techniques contribute significantly to simulating the complex interactions and behaviors observed in real-world traffic scenarios. In the context of mixed traffic studies, these models can be used to create realistic traffic scenarios for various purposes, such as traffic simulation, scenario generation, data augmentation, anomaly detection, policy evaluation, and human behavior modeling.

TABLE 3.2
A comparison of DRL and AIRL

Methods	Advantages	Disadvantages
DRL	*Adaptability*: DRL can adapt to dynamic environments in traffic simulations, by continually updating its policy based on interactions with the environment.	*Weak Robustness:* DRL-based methods can be sensitive to changes in the environment, such as changes in the reward function or the introduction of new obstacles.
	Real-Time Decisions: DRL can make real-time decisions, which is essential for simulating traffic scenarios that require quick responses to changing conditions.	*Sparse Rewards*: In traffic simulations with sparse rewards, DRL might struggle to optimize its policy due to the lack of sufficient feedback.
	Flexibility: DRL supports the exploration of various strategies and can learn optimal policies in different traffic situations.	*Sensitivity to Hyperparameters*: Hyperparameter tuning can be challenging in DRL, especially in complex traffic simulations with multiple interacting agents.
AIRL	*Expert Demonstrations*: AIRL can capitalize on expert demonstrations, including real-world traffic data, to expedite the learning process and generate more accurate traffic simulations.	*Dependence on Demonstrations*: AIRL relies on the availability of expert demonstrations, which may be difficult to obtain or may not cover all possible traffic scenarios.
	Transferability: AIRL learns a reward function, which can potentially make it easier to transfer the learned knowledge to other traffic scenarios or environments.	*Mode Collapse*: AIRL can suffer from mode collapse, where the learned policy may only cover a subset of expert behavior, resulting in less diverse traffic simulations.
	Robustness to Noise: AIRL is more robust to noise in expert demonstrations, making it suitable for traffic simulations where the data might be noisy or incomplete.	*Computational Complexity*: AIRLs adversarial training process for learning both the policy and the reward function can be computationally demanding, especially for large-scale traffic simulations.

Some common types of DGMs include the Deep Autoregressive Model (DAR), Variational Auto-encoder (VAE), Generative Adversarial Network (GAN), and Diffusion Model. Once the DGMs are successfully trained, they will turn into behavioral models and can be deployed to sample new behaviors of BVs. A comparison of these methods when applied in traffic simulation is shown in Table 3.3 (Chen et al., 2024).

3.1.4 Other Deep Learning Models

Playing a crucial role in advancing the field of mixed traffic studies, deep learning models leverage their capability for automated feature learning and complex pattern

TABLE 3.3
A comparison of deep generative models

Methods	Advantages	Disadvantages
DAR	*Temporal Context Sensitivity:* DAR models are capable of incorporating a rich temporal context, which is crucial for modeling time-dependent traffic patterns and predicting future states.	*Computational Intensity:* Implementing and training DAR models can be computationally demanding, particularly for large-scale traffic networks, potentially requiring significant computational resources.
	Non-Linear Relationships: DAR models can capture complex non-linear dependencies within traffic data, allowing for a more accurate representation of dynamic traffic behaviors.	*Model Interpretability:* The high complexity of neural networks may make it more challenging to interpret the learned relationships between variables compared to simpler models.
VAE	*Latent Space Representation:* VAEs provide a compact latent space representation of traffic data, allowing for meaningful feature extraction, anomaly detection, and traffic segmentation.	*Latent Space Ambiguity:* Interpreting the latent space can be challenging, as individual dimensions may not have clear semantic meanings. Understanding the learned representations may require additional effort.
	Generative Capabilities: VAEs can generate synthetic traffic samples from the learned latent space distributions, facilitating the creation of new scenarios for micro-traffic simulation.	*Mode Collapse:* VAEs may struggle to capture the full diversity of complex traffic patterns, potentially leading to the generation of samples that lack variability.
GAN	*Realistic Sample Generation:* GANs excel at producing highly realistic and diverse traffic samples, which is essential for simulating complex and dynamic traffic environments.	*Mode Collapse:* Like VAEs, GANs can suffer from mode collapse, where the model generates a limited range of traffic scenarios, potentially leading to unrealistic simulations.
Diffusion model	*Capturing Complex Dependencies:* Diffusion models excel at modeling intricate dependencies between variables in traffic data, allowing for a more accurate representation of traffic dynamics.	*Interpretability Challenges:* The complex nature of diffusion models may make it more challenging to interpret the learned relationships between variables, requiring additional analytical effort.

recognition. Their versatile application encompasses several key aspects, contributing to a more nuanced understanding of the dynamics within mixed traffic environments.

1. Traffic Flow Prediction: Deep learning models excel in predicting traffic flow patterns by analyzing vast datasets. Utilizing recurrent neural networks (RNNs) (Elman, 1991), Long Short-Term Memory (LSTM) (Gers, Schmidhuber, & Cummins, 2000) networks, or Transformers (Vaswani et al., 2017), these models can capture temporal dependencies and anticipate future traffic conditions. This capability is crucial for proactive traffic management and congestion avoidance.
2. Object Detection and Recognition: Especially in the context of mixed traffic scenarios, deep learning models, such as convolutional neural networks (CNNs) (Fukushima, 1980), facilitate precise object detection and recognition. This includes identifying and tracking various road users, such as vehicles, bicycles, and pedestrians. The application of these models enhances safety and supports the development of AV technologies.
3. Simulation Optimization: By engaging in iterative interactions with the simulated environment, these models learn optimal decision-making policies, resulting in enhanced traffic management and control strategies.
4. Human–Machine Interaction Studies: DL models contribute significantly to the study of human–machine interactions within mixed traffic scenarios. This involves analyzing how drivers and pedestrians interact with AVs or other intelligent traffic systems. Insights gained from these studies inform the development of systems that prioritize safety and harmonious coexistence.

3.2 MODEL-BASED APPROACH

The model-based approach involves scientific hypotheses regarding car movement, considering factors like current position, velocity, acceleration, and road constraints (e.g., intelligent driver model (IDM; Treiber, Hennecke, & Helbing, 2000), Krauss, Adaptive Cruise Control (ACC)). These models are transparent and interpretable, enabling easy understanding and modification of behavior rules by human experts. Moreover, they exhibit computational efficiency, avoiding the need for extensive training data or complex neural network architectures. However, their limitations include a lack of consideration for the anthropomorphic nature of agents and dynamic environmental adaptability, restricting their application in less well-defined situations. Model-based directions typically categorize models into two types: horizontal control models (lane change models) and vertical control models (car-following models).

3.2.1 Car-Following Models

A car-following model describes the movements of a following vehicle (FV) in response to the actions of the lead vehicle (LV). The first car-following models (Pipes, 1953) were proposed in the mid-1950s. A substantial number of models have been developed, including the Gazis–Herman–Rothery model (Gazis, Herman, & Rothery, 1961), the IDM the optimal velocity model (Bando, Hasebe, Nakayama, Shibata, & Sugiyama, 1995), and those proposed by Helly (1959), Gipps (1981), and Wiedemann (1974), among others. For an in-depth review and historical development of the subject, interested individuals should refer to works by Brackstone and McDonald (1999), Olstam and Tapani (2004), Panwai and Dia (2007), and Saifuzzaman and Zheng (2014). In general, car-following models can be categorized into five types: stimulus-based, safety-distance, desired measures, optimal velocity, and psycho-physical models (Zhu, 2022).

1. Stimulus-Based Models: These models dictate that the acceleration of an FV is determined by the driver's reaction to speed and position differences with the vehicle in front. The General Motors models stand out as among the most well-known stimulus-based models, having undergone development since the late 1950s, including one of their recent modifications proposed by Ozaki.
2. Safety-Distance Models: These models operate on the premise that the driver of an FV would adjust speed and maintain a distance, enabling a safe stop if the vehicle in front brakes suddenly. The Gipps model[6] is grounded in the safety-distance concept.
3. Desired Measures Models: These models posit that a driver has a preferred situation defined by specific measures (e.g., following distance and following speed), and the driver consistently endeavors to minimize the disparity between the actual and preferred situations. The IDM is a prominently utilized model within the desired measures category.
4. Optimal Velocity Models: These models presume that each FV possesses an optimal safe velocity contingent on the distance from the lead vehicle. The acceleration of the following vehicle can be ascertained based on the disparity between the actual velocity and the optimal velocity. The full velocity difference (FVD) model (Jiang, Wu, & Zhu, 2001) is closely associated with the optimal velocity model.
5. Psycho-Physical Models: These models propose that a driver's behavior varies based on the traffic state, including whether the driver is in free-flow conditions, approaching the vehicle in front, following the vehicle in front, or braking. The boundary conditions defining these different states are typically articulated as a combination of relative speed and relative distance to the lead vehicle.

3.2.2 Lane Change Models

A lane change, as depicted in Figure 3.1 (Zhu, 2022), is characterized by a purposeful and significant lateral shift in a vehicle's position while moving in the same direction. This maneuver encompasses simple lane changes, merging, exiting, passing, and weaving maneuvers (Chovan, Tijerina, Alexander, & Hendricks, 1994). Lane changes are crucial during automobile journeys and are often linked to traffic incidents (Pande & Abdel-Aty, 2006). In a typical lane change crash scenario, a vehicle intentionally

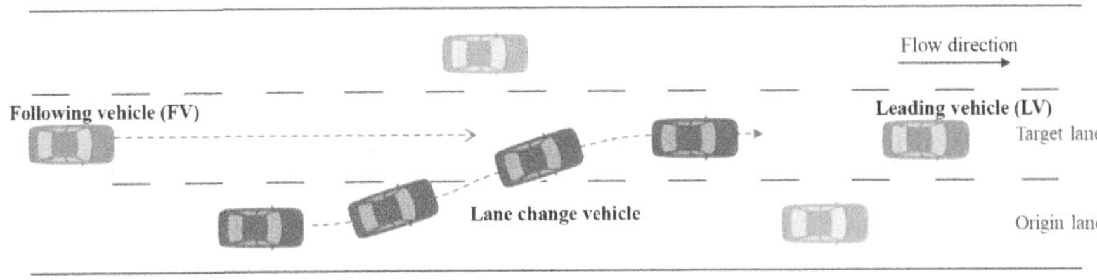

FIGURE 3.1 A figure showing a traffic scenario where a red Lane Change Vehicle moves from the Origin lane to the Target lane between an LV and an FV, both shown in orange.

TABLE 3.4
A comparison of lane change models

Lane-change model	Rule-based	Discrete choice	Incentive-based
Representative models	Gipps model (Gipps, 1986)	Toledo (Toledo, 2003)	MOBIL (Kesting, Treiber, & Helbing, 2007)
Lane-change decision	A series of fixed if-else conditions	Use logit or probit model	Tradeoff of own benefits and negative effects on others
Target lane selection	Lane-change purposes or advantage of changing to a target lane	Utilities for all alternative lanes are calculated and compared	Depend on the level of lane-change desire
Gap acceptance	Gap acceptance parameters are based on simulation or field experiment data	Depending on target lead and lag relative speeds, distance between target lead and lag	Based on deceleration of different alternatives calculated from car-following models
Driver heterogeneity	Not consider driver variations on gap acceptance	Random terms in utility functions	Describe driver variations through politeness factor (MOBIL)
Advantages	Simple in modeling; Decision-making tree with a small number of variables	Having utility theory as a theory basis; Probabilistic results	Small number of parameters; Taking driver variations into consideration
Disadvantages	Hard to calibrate the model parameters; Binary lane-change decisions	Hard to validate and verify the utility function	May not work properly in heavily congested traffic conditions

changes lanes and either sideswipes or is sideswiped by another vehicle proceeding straight in the target lane (Sen, Smith, & Najm, 2003). Furthermore, the adverse impact of lane changes on traffic breakdowns and the reduction of bottleneck discharge rates at the onset of congestion have been researched and validated in prior studies (Cassidy & Rudjanakanoknad, 2005).

Studies related to lane-change models can be categorized into two types: one concentrates on the decision to initiate lane changes and gap acceptance, while the other type centers on the specific operational controls governing lane-change behavior. The majority of lane-change research within the transportation community pertains to the first type (decision-making). The second type of lane-change study is typically associated with automation and driving assistance systems and is conducted by individuals in the automation community.

Lane-change models encompass two types of studies: one concentrates on initiating lane changes and gap acceptance decisions, while the other focuses on the operational controls governing lane change behavior. The majority of lane-change research within the transportation community pertains to the first type (decision-making). The second type of lane-change study is typically associated with automation and driving assistance systems and is conducted by individuals in the automation community. This section provides an overview of the most widely recognized lane-change models developed in the transportation community.

As summarized in Table 3.4, models describing lane change decisions can be categorized into the following types. Detailed reviews of these models are available in Rahman, Chowdhury, Xie, and He (2013), Moridpour, Sarvi, and Rose (2010), and Yunus, Razak, Yogarayan, and Abdullah (2021).

3.3 AVAILABLE OPEN DATASETS

In mixed traffic research, the quality and richness of data play a crucial role in model training. The quantity and quality of information provided to the model have a significant impact on the outcomes it produces. We introduce the driving dataset with different views, Field of View (FOV) and Bird's Eye View (BEV). FOV datasets are better suited for training AV algorithms that rely on egocentric information, while BEV datasets are limited to a fixed region, making them more suitable for analyzing the behavior of objects against a stable background. Then, the datasets are summarized based on their data modalities, including camera, LiDAR, GNSS, and HD map. We compared the volume of datasets based on the number of scenes, hours, and frames. Weather conditions can also affect the performance of the model. Some datasets, such as the Waymo open dataset, encompass all weather conditions, ranging from sunny to snowy. The details are provided in Table 3.5 (Chen et al., 2024). It is worth mentioning that the year in the table refers to the most recent year in which the dataset was updated.

TABLE 3.5
A comparison of driving datasets

			Data modalities					Data volume			Annotation			
Dataset	Year	View	Camera	LiDAR	Radar	GNSS	HD map	Scenes	Hours	Frames	3D	2D	Lane	Weather
Waymo (Motion)	2023	BEV	√	√	×	√	√	1K	–	15M	√	×	√	×
Argoverse 2 (LiDAR)	2022	BEV	×	×	×	√	√	25M	763	–	×	×	×	×
Argoverse 2 (Motion)	2022	BEV	√	×	×	√	×	3	6	–	√	×	√	×
InD	2020	BEV	√	×	×	√	×	4	16.5	–	×	√	√	×
RounD	2020	BEV	×	×	×	√	√	4	2.5	–	×	√	×	×
INTERACTION	2019	BEV	√	√	√	√	√	366	1118	4.6M	×	√	√	×
Lyft Level 5	2019	BEV	√	√	×	√	×	–	214	2000M	√	√	×	×
HighD	2018	BEV	√	√	×	√	×	–	104	–	√	√	√	√
NGSIM	2007	BEV	√	√	×	√	√	2030	11.3	39M	√	√	√	×
Waymo (Perception)	2023	FOV	×	√	×	√	√	2M	–	6M	×	×	×	√
Argoverse 2 (Sensor)	2022	FOV	√	√	×	√	×	1K	6.4	20M	√	√	√	√
OpenLane	2022	FOV	√	×	×	√	×	–	2.5	5K	√	√	√	×
ONCE-3DLanes	2022	FOV	√	√	×	√	×	179	–	1.4M	√	×	×	√
PandaSet	2021	FOV	√	√	×	√	×	–	144	41.7M	√	√	×	×
ONCE	2021	FOV	√	√	×	×	×	–	–	21.1M	√	√	√	√
A2D2	2020	FOV	√	√	×	√	×	–	–	1.25M	√	√	√	√
Cityscapes 3D	2020	FOV	√	√	×	√	×	11	–	8M	√	√	√	×
KITTI-360	2020	FOV	√	×	×	√	×	4	10	–	√	×	√	×
HDD	2019	FOV	√	√	×	√	√	113	320	2.2M	√	×	√	×
Argoverse 1	2019	FOV	√	√	√	√	√	1K	5.5	4M	√	√	×	×
nuScenes	2019	FOV	×	√	×	√	√	–	574	20M	√	√	√	√
Oxford RobotCar	2019	FOV	√	√	×	√	×	160	0.8	2.8M	√	×	×	√
H3D	2019	FOV	√	√	×	×	×	–	55	3.9M	√	√	×	×
A*3D	2019	FOV	√	√	×	√	√	103	2.5	14.4M	√	√	√	√
ApolloScape	2018	FOV	√	×	×	√	×	–	1100	10M	×	√	√	√
BDD100K	2018	FOV	√	×	×	√	×	6	11.5	–	×	√	√	√
KITTI	2013	FOV	√	√	×	√	×	22	1.5	1.5M	√	√	√	√

REFERENCES

Bando, M., Hasebe, K., Nakayama, A., Shibata, A., & Sugiyama, Y. (1995). Dynamical model of traffic congestion and numerical simulation. *Physical Review E*, 51(2), 1035.

Brackstone, M., & McDonald, M. (1999). Car-following: A historical review. *Transportation Research Part F: Traffic Psychology and Behaviour*, 2(4), 181–196.

Cassidy, M. J., & Rudjanakanoknad, J. (2005). Increasing the capacity of an isolated merge by metering its on-ramp. *Transportation Research Part B: Methodological*, 39(10), 896–913.

Chen, D., Zhu, M., Yang, H., Wang, X., & Wang, Y. (2024). Data-driven traffic simulation: A comprehensive review. *IEEE Transactions on Intelligent Vehicles*, 9(4), 4730–4748. doi:10.1109/TIV.2024.3367919

Chovan, J. D., Tijerina, L., Alexander, G., & Hendricks, D. L. (1994). Examination of lane change crashes and potential IVHS countermeasures. Joint Program Office for Intelligent Transportation Systems. Report Number: DOT HS 808 071;DOT-VNTSC-NHTSA-93-2.

Cui, Y., Huang, S., Zhong, J., Liu, Z., Wang, Y., Sun, C., … Khajepour, A. (2024). DriveLLM: Charting the path toward full autonomous driving with large language models. *IEEE Transactions on Intelligent Vehicles*, 9(1), 1450–1464.

Elman, J. L. (1991). Distributed representations, simple recurrent networks, and grammatical structure. *Machine Learning*, 7, 195–225.

Fukushima, K. (1980). Neocognitron: A self-organizing neural network model for a mechanism of pattern recognition unaffected by shift in position. *Biological Cybernetics*, 36(4), 193–202.

Gazis, D. C., Herman, R., & Rothery, R. W. (1961). Nonlinear follow-the-leader models of traffic flow. *Operations Research*, 9(4), 545–567.

Gers, F. A., Schmidhuber, J., & Cummins, F. (2000). Learning to forget: Continual prediction with LSTM. *Neural Computation*, 12(10), 2451–2471.

Gipps, P. G. (1981). A behavioural car-following model for computer simulation. *Transportation Research Part B: Methodological*, 15(2), 105–111.

Gipps, P. G. (1986). A model for the structure of lane-changing decisions. *Transportation Research Part B: Methodological*, 20(5), 403–414.

Griffith, S., Subramanian, K., Scholz, J., Isbell, C. L., & Thomaz, A. L. (2013). Policy shaping: Integrating human feedback with reinforcement learning. *NIPS'13: Proceedings of the 26th International Conference on Neural Information Processing Systems – Volume 2*. Lake Tahoe, Nevada.

Helly, W. (1959). *Simulation of Bottlenecks in Single-Lane Traffic Flow*. In *Proceedings of the Symposium on Theory of Traffic Flow* (pp. 207–238). Research Laboratories, General Motors.

Jiang, R., Wu, Q., & Zhu, Z. (2001). Full velocity difference model for a car-following theory. *Physical Review E*, 64(1), 017101.

Kesting, A., Treiber, M., & Helbing, D. (2007). General lane-changing model MOBIL for car-following models. *Transportation Research Record*, 1999(1), 86–94.

Moridpour, S., Sarvi, M., & Rose, G. (2010). Lane changing models: A critical review. *Transportation Letters*, 2(3), 157–173.

Olstam, J. J., & Tapani, A. (2004). *Comparison of Car-Following Models* (Vol. 960). Sweden: Swedish National Road and Transport Research Institute Linköping.

Pande, A., & Abdel-Aty, M. (2006). Assessment of freeway traffic parameters leading to lane-change related collisions. *Accident Analysis & Prevention*, 38(5), 936–948.

Panwai, S., & Dia, H. (2007). Neural agent car-following models. *IEEE Transactions on Intelligent Transportation Systems*, 8(1), 60–70.

Pipes, L. A. (1953). An operational analysis of traffic dynamics. *Journal of Applied Physics*, 24(3), 274–281.

Rahman, M., Chowdhury, M., Xie, Y., & He, Y. (2013). Review of microscopic lane-changing models and future research opportunities. *IEEE Transactions on Intelligent Transportation Systems*, 14(4), 1942–1956.

Saifuzzaman, M., & Zheng, Z. (2014). Incorporating human-factors in car-following models: A review of recent developments and research needs. *Transportation Research Part C: Emerging Technologies*, 48, 379–403.

Sen, B., Smith, J. D., & Najm, W. G. (2003). Analysis of lane change crashes. No. HS-809 571. Department of Transportation. National Highway Traffic Safety Administration.

Toledo, T. (2003). *Integrating Driving Behavior Modeling*. Massachusetts Institute of Technology.

Treiber, M., Hennecke, A., & Helbing, D. (2000). Congested traffic states in empirical observations and microscopic simulations. *Physical Review E*, 62(2), 1805.

Vaswani, A., Shazeer, N., Parmar, N., Uszkoreit, J., Jones, L., Gomez, A. N., … Polosukhin, I. (2017). Attention is all you need. *Advances in Neural Information Processing Systems*, 30.

Wiedemann, R. (1974). *Simulation des strabenverkehrsflusses in schriftenreihe des tnstituts fir verkehrswesen der universitiit karlsruhe*.Karlsruhe, Germany: Universität Karlsruhe.

Yunus, N. E., Razak, S. F. A., Yogarayan, S., & Abdullah, M. F. A. (2021). Lane changing models: A short review. *Paper presented at the 2021 IEEE 12th Control and System Graduate Research Colloquium (ICSGRC)* (pp. 110–115). Shah Alam, Malaysia.

Zhu, M. (2022). *Behavior Modeling and Motion Planning for Autonomous Driving using Artificial Intelligence*. Doctoral dissertation.

4 Micro–Macro Mixed Traffic Flow Modeling

Jian Rong, Yi Wang, Chenjing Zhou, and Yue Chen

4.1 INTRODUCTION

With urbanization accelerating, urban populations concentrate and expand, exacerbating the gap between transportation supply and demand. Traffic congestion, now prevalent even in smaller cities, brings safety, pollution, noise, and energy consumption issues. These challenges hinder urban economic efficiency, thereby stalling sustainable development (Lu et al., 2014).

Since the first car in 1866, automobiles have become indispensable, boosting the automotive industry. By 2015, the global vehicle fleet reached 1.282 billion, a 43.72% increase from 2005, with congestion causing significant societal and economic damage. Predictions suggest congestion costs in Europe and the US could hit $293.1 billion by 2030. Each household's annual congestion cost could reach $3,217 in the UK and $2,301 in the US. This underscores the urgency for congestion management to enhance traffic flow.

Two main methods alleviate congestion: capacity enhancement through design and delay reduction via optimized control schemes. Microscopic traffic flow models analyze driving behavior's impact on vehicle speed, affecting macroscopic traffic flow, headway, and density. This chapter explores traditional microscopic car-following and lane-changing models (Sections 4.2 and 4.3), traditional macroscopic models (Section 4.4), dynamics, sensors, and traffic flow simulation tools (Section 4.5).

4.2 CAR-FOLLOWING MODEL

In the middle of the 20th century, Pipes first proposed the concept of vehicle following based on the idea that the two neighboring vehicles in front and behind should maintain a safe-following distance and constructed the kinetic trajectory equations of the subject vehicle through the following distance (Pipes, 1953). Vehicle-following behavior, the most basic microscopic driving behavior, refers to the interaction between the front and rear vehicles on a single lane where overtaking is not allowed (see Figure 4.1). The following model is a mathematical description of the rear vehicle adjusting its own motion state according to the motion state of the vehicle in front of

 DOI: 10.1201/9781032674667-4

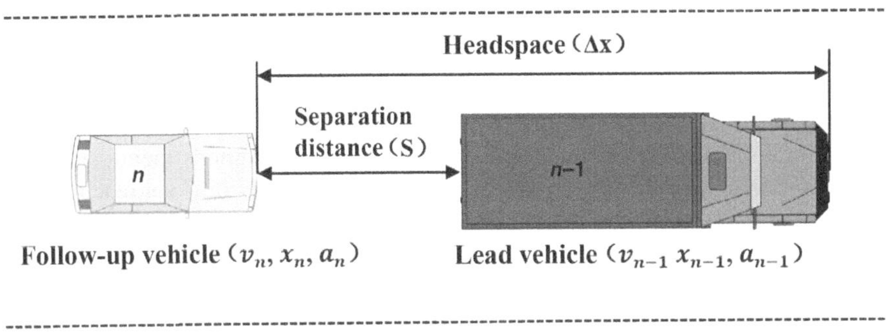

FIGURE 4.1 Diagram of car-following behavior.

it (Wang et al., 2011), which is the theoretical basis of the capacity estimation, forward collision avoidance, and adaptive cruise control system, and also the core model of the microscopic traffic simulation software. The following model analyzes the way of following between vehicles to understand the traffic flow characteristics of a single lane, and plays the role of linking the microscopic driving behavior and macroscopic traffic flow (Wang & Jin, 2012).

4.2.1 Traditional Car-Following Model

The traditional heeling models that are widely used at present are classical models that have been tested for a long time, so the classical models (e.g., **Gazis-Herman-Rothery** (GHR) model, Gipps model, Full Velocity Difference (FVD) model, Wiedemann model, and Intelligent Driver Model (IDM)) in the stimulus–response category, safety distance category, intelligent driver model, psycho-physiological category, and optimized speed category are introduced below to be investigated. The introduction of relevant models is covered in Chapter 3.

4.2.2 Autonomous Driving Heeling Models

The mainstream approaches in modeling self-driving follower models can be classified into four categories: rule-based methods, supervised learning methods, model predictive control methods, and reinforcement learning methods.

Rule-based methods mainly use the traditional heeling models or improve such models. For example, Zhou et al. (2016) used the IDM model in Adaptive Cruise Control (ACC). Supervised learning approaches use machine learning methods to summarize the relationship between vehicle states and acceleration changes from human driving data to build a longitudinal model of the vehicle (Kuefler et al., 2017). Zhou et al. (2016) used recurrent neural networks to learn the "stop-and-go driving behaviors" of a human driver in a traffic-oscillating environment, which is a much more effective approach to learning the "stop-and-go driving behaviors" of a human driver in a traffic-oscillating environment, as compared to the traditional IDM model and the artificial driving model. Huang et al. (2018) used long-short-term memory (LSTM) network to model asymmetric driving behavior and achieved better results than recurrent neural networks. Ma & Qu (2020) used sequence-to-sequence (seq2seq) network to build a multi-step output model, which can significantly reduce the trajectory error in multi-vehicle driving scenarios in the traffic shock phase. This type of supervised learning-based method can better fit human driving behavior than a traditional following model. However, these two methods mentioned above will incorporate the inherent defects of human driving behavior (long reaction time, heterogeneity, etc.) into the model during the process of imitating a human driver's following behavior. At the same time, the model is not flexible enough to add new optimization objectives, so these two methods cannot be used as the optimal solution for the longitudinal control strategy of autonomous driving (Chai & Wong, 2015; Basu et al., 2017).

Compared with the drawbacks of the previous two approaches, the model-based predictive control approach is more flexible and can consider more optimization objectives and constraints, which has become the mainstream research method for longitudinal control modeling of autonomous driving in the past (Wegener et al., 2021). This method provides driving decisions by predicting the future state of the trajectory of the vehicle in front of it, which improves the optimization and control performance (Zhou et al., 2019). Luo et al. (2010) came up with a model predictive control strategy for real driving scenarios, which tries to control the acceleration of the following vehicle to keep the relative distance between the two vehicles within a safe range. Simulation results show that this control strategy can improve the safety of the vehicle during driving compared to a human driver. An ACC model based on model predictive control was also proposed to solve the problems of following, fuel consumption and driver expectation response (Takahama & Akasaka, 2018). The results showed that the developed ACC model achieved satisfying results in terms of fuel utilization, following performance, and driver expectations. However, the model prediction approach usually requires high computational power and is not applicable to real-time control.

The development of reinforcement learning in recent years has brought new solution ideas to sequential decision-making problems such as autonomous driving and has been widely used in longitudinal control research for autonomous driving (Karnchanachari et al., 2020). Reinforcement learning has two main advantages over model predictive control. First, in the testing process, the reinforcement learning control strategy only needs to take the state as input and output the corresponding action, while the model predictive control needs to solve the finite time-constrained optimal control problem in each time step, so the reinforcement learning control strategy has lower arithmetic power requirement in comparison with the application (Ernst et al., 2008). Second, some studies have shown that reinforcement learning exhibits better performance relative to traditional model predictive control (Lin et al., 2020). Zhu et al. (2019) used a deep reinforcement learning algorithm to construct a vehicle-following model and used real vehicle data to verify the effectiveness of the model. In another study, based on the driving characteristics of human driving data, a reinforcement learning-based heeling model was constructed (Pan & Bao, 2019). Zhu et al. (2020) summarized a set of methods for evaluating efficiency by analyzing the driving data of natural drivers. Then, they proposed a reward function that considered three optimization objectives, namely safety, comfort, and efficiency, and used a deep reinforcement learning algorithm to achieve better results compared to model predictive control.

4.3 LANE-CHANGING MODEL

4.3.1 Lane Change Models in Conventional Scenarios

Vehicle lane-changing behavior is a common driving behavior in which drivers, often dissatisfied with the current driving speed on the road, pursue a higher driving speed or more adequate driving space and thus take lane-changing action. The lane-changing behavior of a vehicle can be divided into three stages: the generation of lane-changing motivation, the judgment of lane-changing conditions, and the execution of lane-changing decisions. Nowadays, the mainstream lane-changing models can be roughly divided into four categories: rule-based lane-changing models, discrete choice lane-changing models, incentive-based lane-changing models, game theory-based lane-changing models. The introduction of relevant models is covered in Chapter 3.

4.3.2 Lane-Changing Models for Hybrid Traffic Streams

Currently, most of the research on vehicle lane-changing models in human–machine mixed driving traffic flow starts from the two aspects of lane-changing intention and lane-changing decision-making, and constructs lane-changing models by considering the behavioral variability of different vehicles. The research on lane-changing intention mainly considers the driving conditions under which the vehicle generates lane-changing intention, and the lane-changing intention is generated when the vehicle speed is limited; for the research on lane-changing decision-making, researchers mainly consider the relative speeds of the lane-changing vehicle and the surrounding vehicles and the spatial conditions of the target lanes, etc., and construct the lane-changing decision-making model based on the behavioral variability of different types of vehicles to describe the lane-changing behavior of the vehicles in the hybrid traffic flow. For example, Zou & Yang (2002) considered that vehicles are restricted by the speed of the vehicle in front of them to generate lane-changing intentions; Yang & Koutsopoulos (1996) considered different spatial conditions affecting vehicle lane-changing behavior and established a lane-changing model based on the expectation model and the gap acceptance theory to consider whether a vehicle generates lane changing intentions. Chen & Wang (2019) established a cumulative model of driver dissatisfaction based on the driver's desired speed and generated lane-changing intentions based on the cumulative degree of driver dissatisfaction with the operating speed. Liu et al. (2017) analyzed the characteristics of human–machine-mixed-driving traffic flow, and classified the vehicle lane-changing rules into two categories, that is, courtesy-type lane-changing and aggressive-type lane-changing: all the vehicles can follow the aggressive lane-changing rules, but only the autonomous vehicles follow the courtesy-type lane-changing rules. Tanveer et al. (2020) considered basic traffic parameters and driver characteristics to classify self-driving vehicles and manually driven vehicles into two sub-classes and constructed a lane-changing model for each sub-class to simulate the impact of self-driving vehicles on mixed traffic flow. As summarized in Table 4.1, these studies categorize lane-changing models based

TABLE 4.1
Summary of mixed human–machine traffic flow lane-changing models

Research area			Related literature	Main research contents
Lane-changing intention			Zou & Yang, 2002	Consider a vehicle limited by the speed of the vehicle in front of it
			Yang & Koutsopoulos, 1996	Consider the impact of different spatial conditions
			Chen & Wang, 2019	Consider driver dissatisfaction
Lane-changing decision	Lane-changing rules		Liu et al., 2017	Classifying vehicle lane-changing rules into two categories: polite and aggressive lane-changing
			Tanveer et al., 2020	Classification of self-driving vehicles and manually driven vehicles by considering basic traffic parameters and driver characteristics
	Lane-changing behavior	Coordinated lane-changing	Li et al., 2020; Luo et al., 2019	Consider the case of cooperative lane-changing between independent CAVs.
			Ni et al., 2020	A multi-objective optimization control function for multi-lane cooperative lane-changing is established, and a two-stage cooperative lane-changing framework is proposed
		Vehicle queue	Sun et al., 2021	Consider overall lane-changing of vehicle queues
			Liu et al., 2018	Consider a scenario where networked self-driving vehicles dynamically join the vehicle queue
			An & Talebpour, 2019	The impact of lane-changing behavior on cars in the vehicle queue is investigated from the perspective of optimized lane-changing trajectories
	Model parameters	Basic parameters	Yang et al., 2014	Parameters such as the vehicle's desired deceleration were calibrated from the five major lanes of U.S. Highway 101 in the southbound direction from the NGSIM trajectory data
			Wang et al., 2020	Parameter calibration and validation of common multiple lane-changing models through natural highway driving data of 48 drivers
		Special parameters	Zhou & Zhu, 2021	For the maximum queue length, an analytical method regarding road capacity and traffic flow stability is proposed
			Liu et al., 2018; Luo et al., 2018	Consider capacity and mobility to explore maximum queue lengths
			Liu et al., 2018; Zhou et al., 2020	Consider net-CAVs performing immediate lane-changing behavior when safety conditions are met

on lane-changing intention, rules, and coordinated lane-changing in mixed human–machine traffic flow.

In the human–machine mixed-driving traffic flow, the unique vehicle queuing system of networked self-driving vehicles will derive more vehicle lane-changing behaviors, such as cooperative lane-changing behaviors based on vehicle–vehicle communication, and lane-changing behaviors such as dynamic vehicle convergence or departure from the queue due to the decomposition and reorganization of the vehicle queue. For example, Li et al. (2020) and Luo et al. (2019) constructed a lane-changing model by considering the cooperative lane-changing situation among independent network-connected autonomous vehicles (CAVs). Ni et al. (2020) established a multi-objective optimization control function for cooperative lane-changing of multiple lanes and proposed a two-stage cooperative lane-changing framework, which divides the lane-changing process into a lane-changing phase and a longitudinal vehicle distance adjustment phase. As for vehicle queues, Sun et al. (2021) constructed a lane-changing model considering the overall lane-changing situation of vehicle queues and Liu et al. (2018) considered the scenario of grid-CAVs dynamically joining the vehicle queues when studying the vehicle queue system. The impact of lane-changing behavior on the vehicle queues from the perspective of the optimal lane-changing trajectory was also considered in An & Talebpour (2019).

In the existing mixed-driving traffic flow research regarding lane-changing behaviors, the selection of relevant parameters for manually driven vehicles generally follows the traditional traffic flow calibration results, and the research on relevant parameters for networked self-driving vehicles can be roughly divided into two aspects: basic parameters and unique parameters. In terms of basic parameters, most scholars followed the same performance parameters as manually driven vehicles, such as vehicle acceleration and deceleration and only a few scholars tried to determine the values of parameters affecting lane-changing behavior, such as the maximum queue length of the grid-CAVs, to optimize the macroscopic communication characteristics. For example, Yang et al. (2014) calibrated the parameters such as the expected vehicle deceleration speed through the southbound direction data of five major lanes of U.S. Highway 101 in the NGSIM trajectory data; Wang et al. (2020) calibrated the parameters and verified the effect of a variety of common lane-changing models through the freeway natural driving data of 48 drivers, and so on. In terms of vehicle queue length, Zhou & Zhu (2021) suggest that the maximum queue length of networked self-driving vehicles should be no more than 10 vehicles by analyzing the stability of traffic flow, while Liu et al. (2018) and Luo et al. (2018) suggest that the maximum queue length should be 10–20 vehicles based on the considerations of capacity and mobility.

4.4 MACRO TRAFFIC FLOW MODELING

4.4.1 Traditional Models

4.4.1.1 Fundamental Diagram

The first speed–density relationship in the history of traffic flow theory research was proposed by Greenshields et al. (1935) when they studied the U.S. highway traffic

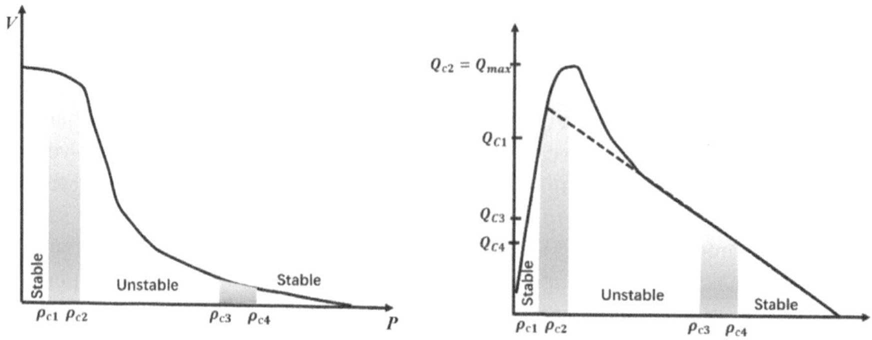

FIGURE 4.2 Schematic of fundamental diagram in traffic flow theory. The left diagram shows the speed–density relationship, while the right diagram illustrates the flow–density relationship.

flow, and Greenshields gave the speed–density relationship in the form of $v_f\left(1-\frac{k}{k_j}\right)$, where v_f and k_j are the maximum speed of the chimney by the flow state and the maximum density of the blockage state, respectively. The above equation is known as the linear equilibrium speed–density relationship and is still widely used today (see Figure 4.2).

Until the macroscopic continuum and microscopic vehicle following models proposed in the 1950s–70s, the gas-kinetic theoretical models proposed in the 1970s–90s, and the metric automata models from the 1990s to the present, the research work on theoretical modeling of traffic flow has flourished. However, almost all these theoretical models are based on the same basic assumption, that is, that the stationary solution of the model belongs to a curve in the flow–density plane. This curve passes through the origin of the flow–velocity diagram and has at least one extreme value. This curve is called the "Fundamental Diagram of the traffic flow" (Schönhof & Helbing, 2007). In other words, the above linear fundamental diagram of traffic flow reflects an important assumption of the traditional traffic flow model, that is, there is a single-valued correspondence between flow and density, and there can be only one steady-state flow rate at a density, and the flow rate is a single-valued function of the density, which responds to the model's equilibrium flow–density or speed–density relationship. The theoretical modeling approach based on the corresponding assumptions is commonly known as the "Fundamental Diagram Approach."

4.4.1.2 Three-Phase Traffic Model

Due to the limitations of the fundamental diagram approach, in the years 1996–2004, Kerner (1996, 1997, 1998, 1999, 2000, 2004) developed a Three-Phase Traffic Theory based on extensive real-world data from highways. During this period, Kerner and his collaborators conducted numerous studies, gradually establishing the foundations of the Three-Phase Traffic Theory through the analysis of real-world data and theoretical modeling. Kerner (2004) published his seminal work "The Physics of Traffic," marking the formation of a comprehensive theoretical framework for the Three-Phase Traffic Theory. This emerging theory in traffic flow attracted widespread attention from physicists and traffic experts, leading to a plethora of research endeavors centered around this new theoretical framework.

The Three-Phase Traffic Theory divides traffic states into three phases: Free flow (F), Synchronized flow (S), and Wide Moving Jam (J). Consequently, there exist three phase transitions theoretically: from Free flow to Synchronized flow (F–S), from Free flow to Jam (F–J), and from Synchronized flow to Jam (S–J). Kerner & Rehborn (1997) pointed out that in real traffic processes, all three phase transitions can occur and manifest as first-order phase transitions. Kerner further emphasized that the most common phase transitions occur between free flow and synchronized flow, as well as between synchronized flow and jamming, which are the first and third transitions mentioned above (Kerner, 1998). Particularly, the spontaneous formation of wide moving jams within free flow on uniform roads typically involves two first-order phase transitions: first, a transition from free flow to synchronized flow, followed by a transition from synchronized flow to jamming, resulting in wide moving jams forming within synchronized flow.

4.4.1.3 Lighthill–Whitham–Richards (LWR) Model

The traffic flow models can describe many real-world traffic phenomena, but existing classical traffic flow models have some shortcomings when integrating with actual traffic data, which renders them impractical in real-world applications. The first-order LWR model, using the Greenshields speed–density relationship, fails to fully describe the distribution characteristics of speed–density data. Supported by empirical data, this chapter proposes a more accurate fitting method based on the least squares method to fit flow–density data, allowing the fitting results to more fully represent the road information contained in the empirical data.

The LWR model is a macroscopic traffic flow model that describes the propagation characteristics of traffic waves (Lighthill & Whitham, 1955; Richards, 1956). In the LWR model, the characteristic lines of the wave solution correspond to the tangent lines at the equilibrium state of the fundamental diagram. In other words, the propagation speed of traffic waves within a platoon should be equal to the slope of the corresponding $q - k$ fundamental diagram. Therefore, it is necessary to demonstrate that the slope of the tangent line on the mixed traffic flow fundamental diagram, denoted as $\frac{dp}{dk}$, is equal to the wave speed of the traffic waves within the mixed traffic flow. This should be supplemented with numerical simulations for verification. Subsequently, based on the mixed traffic flow fundamental diagram model, an LWR model for the mixed traffic flow is established (see Figure 4.3).

4.4.2 Mixed Traffic Flow Modeling

4.4.2.1 Fundamental Diagram Model Applied in Mixed Traffic Flow

The vehicle following model is an important theoretical tool for the study of macroscopic traffic flow characteristics from a microscopic point of view. The basic

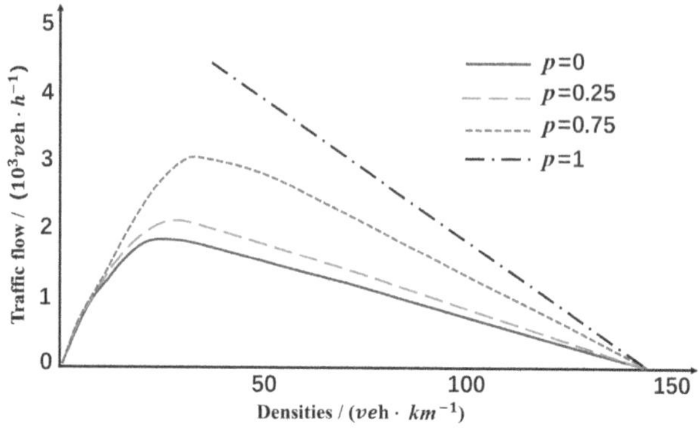

FIGURE 4.3 Basic diagram of mixed traffic flow.

graphical model of traditional traffic flow is generally derived based on the vehicle following distance at the steady state speed of the traffic flow, but in the networked environment where the CAV penetration rate has not reached 100%, the following distance at the steady state is not the same due to the differences in the following behaviors of the front and rear vehicles of CAV-CAV, HV-CAV, CAV-HV, and HV-HV in the mixed traffic flow.

In the following equations, l_{CACC}, l_{ACC}, and l_{HDV} are used to denote the steady state following the distance of the CACC following the relationship, the steady state following the distance of the ACC following the relationship, and the steady state following the distance of the conventional vehicle following the relationship, respectively. From the construction principle of the interaction potential-based vehicle following model, l_{CACC}, l_{ACC}, and l_{HDV} correspond to the demand following distance of the CACC following model based on the interaction potential, the safe following distance of the ACC following model based on the interaction potential, and the safe following distance of the traditional vehicle following model based on the interaction potential, respectively, and can be obtained by putting the steady-state speed versus into the relation equation:

$$\begin{cases} l_{CACC} = s_0 + \beta_{CAV^{V_s}} + \eta_{CACC} \dfrac{V_s^2}{2d_{Fm}} \\ l_{ACC} = s_0 + \beta_{CAV^{V_s}} + \eta_{ACC} \dfrac{V_s^2}{2d_{Fm}} \\ l_{HDV} = s_0 + \beta_{HV^{V_s}} + \eta_{CACC} \dfrac{V_s^2}{2d_{Fm}} \end{cases} \tag{4.1}$$

In a mixed traffic flow consisting of a sufficient number of vehicles, with P_{CACC}, P_{ACC}, and P_{HDV} denoting the proportion of the three following relationships of CACC,

ACC, and HDV, respectively, the longitudinal length of the traffic flow is when all the vehicles are running at a steady-state speed versus:

$$D = Z\left[P_{CACC}\left(l_{CACC} + l_v\right) + P_{ACC}\left(l_{ACC} + l_v\right) + P_{HDV}\left(l_{HDV} + l_v\right)\right] \tag{4.2}$$

where D is the total length of the mixed traffic flow; Z is the total number of vehicles of various types in the mixed traffic flow; and l_V is the longitudinal length of the body.

As a result, the vehicle density k of the mixed traffic flow can be obtained by conversion:

$$k = \frac{1}{\left[P_{CACC}\left(l_{CACC} + l_v\right) + P_{ACC}\left(l_{ACC} + l_v\right) + P_{HDV}\left(l_{HDV} + l_v\right)\right]} \tag{4.3}$$

It is easy to see that the vehicle density k is a function of the various following relationship ratios P_{CACC}, P_{ACC}, and P_{HDV} with the following inter-flow distances l_{CACC}, l_{ACC}, and l_{HDV} in the steady state following relationship as the independent variables, whereas from Eq. (4.1), l_{CACC}, l_{ACC}, and l_{HDV} themselves are functions of the steady-state speed versus as the independent variables, respectively. Therefore, to study the relationship between vehicle density k and steady-state speed versus, we must first analyze P_{CACC}, P_{ACC}, and P_{HDV} in the mixed traffic flow.

In a mixed traffic flow, the arrangement of HVs and CAVs in the longitudinal direction is random. However, when the total number of vehicles Z is large enough, as the CAV occupancy rate p increases in the mixed traffic flow, the proportion of neighboring vehicles in the traffic flow that establish different following relationships will change. The CAV and HV occupancy in the mixed traffic flow are denoted by p and $1 - p$, respectively. According to the principle of probability theory, the attributes of the front and rear vehicles in the following relationship are randomly determined according to their occupancy. According to the principle of probability theory, the attributes of the front and rear vehicles in the following relationship are determined randomly according to their occupancy rates. When the front and rear vehicles are both CAVs, the proportion of the two vehicles to establish CACC following relationship is p^2; when the front and rear vehicles are both HVs, the proportion of the two vehicles to establish HDV following relationship is $(1 - p)^2$; when HVs are following the CAVs, the proportion of the two vehicles to establish HDV following relationship is $p\,(1 - p)$; and when the CAVs are following the HVs, the proportion of the two vehicles to establish ACC following relationship is $(1 - p)p$.

P_{CACC}, P_{ACC}, and P_{HDV} formulas:

$$\begin{cases} P_{HDV} = 1 - p \\ P_{ACC} = (1 - p)\,p \\ P_{CACC} = p^2 \end{cases} \tag{4.4}$$

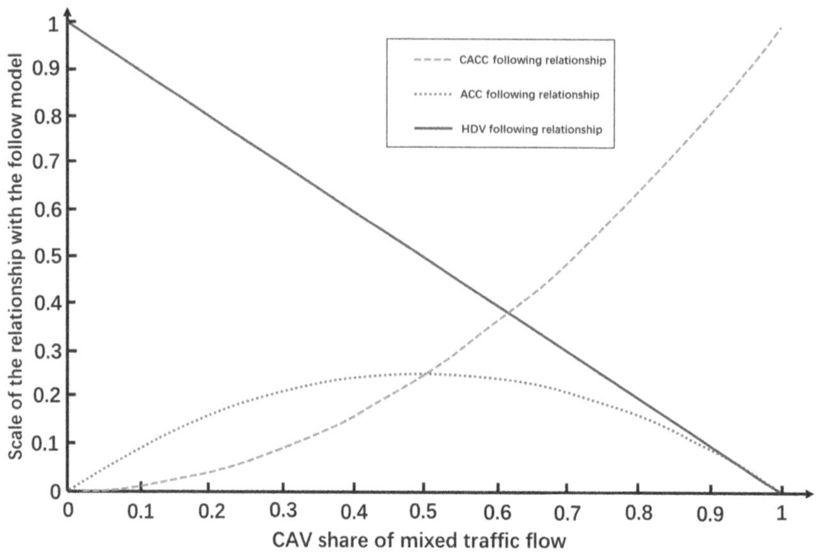

FIGURE 4.4 Relationship between car following proportion and CAV proportion.

The relationship between the three types of following ratios and the CAV's share in the mixed traffic flow is calculated and shown in Figure 4.4. As the proportion of CAV increases: the proportion of HDV following relationship decreases linearly; the proportion of ACC following relationship increases and then decreases, and is the highest at $p = 0.5$; the proportion of CACC following relationship increases with a quadratic trend.

4.4.2.2 LWR Model Applied in Mixed Traffic Flow

The LWR model was proposed by Lighthill et al. in the middle of the 20th century (Lighthill & Whitham, 1955), and has been widely used for analyzing the macroscopic characteristics of traffic flow. According to the principle of the LWR model, the velocity of the oscillating wave of the traffic flow relative to the road surface is the partial differential of the steady-state traffic flow concerning the vehicle density, and its expression is:

$$V_W = \frac{dq}{dk} \tag{4.5}$$

where V_W is the relative roadway velocity of the traffic shock wave. When it is positive, the shock wave propagates upstream of the roadway; and when it is negative, it propagates downstream of the roadway. The Eq. 4.5 is deformed as:

$$V_W = \frac{dq}{dV_S} \cdot \frac{dV_S}{dk} \tag{4.6}$$

Calculating the partial differential of the flow rate with respect to the velocity and simplifying, we can obtain:

$$\frac{dq}{dV_S} = k - k^2 V_S \left[(1-p)\left(\beta_{HV} + \frac{V_S}{d_{Fm}}\right) + p\left(\beta_{CAV} + \frac{\eta_{ACC^{V_S}}}{d_{Fm}}\right) + \frac{p^2 V_S (\eta_{CACC} - \eta_{ACC})}{d_{Fm}} \right] \tag{4.7}$$

Calculating the partial differential of velocity with respect to density and simplifying, we can obtain:

$$\frac{dV_S}{dk} = \frac{-1}{k^2 \left[(1-p)\left(\beta_{HV} + \frac{V_S}{d_{Fm}}\right) + p\left(\beta_{CAV} + \frac{\eta_{ACC^{V_S}}}{d_{Fm}}\right) + \frac{p^2 V_S (\eta_{CACC} - \eta_{ACC})}{d_{Fm}} \right]} \tag{4.8}$$

Substituting Eq. (4.7) and Eq. (4.8) back into Eq. (4.6), we can get:

$$V_W = V_S - \frac{V_S^2 \left[\frac{p^2(\eta_{CACC} - \eta_{ACC}) + p(\eta_{ACC} - 1) + 1}{2d_{Fm}} + \frac{(1-p)\beta_{HV} + p\beta_{CAV}}{V_S} \right] + S_0 + l_V}{(1-p)\left(\beta_{HV} + \frac{V_S}{d_{Fm}}\right) + p\left(\beta_{CAV} + \frac{\eta_{ACC^{V_S}}}{d_{Fm}}\right) + \frac{p^2 V_S (\eta_{CACC} - \eta_{ACC})}{d_{Fm}}} \tag{4.9}$$

It is easy to see that the speed of the traffic shock wave in the mixed traffic flow is a function of the CAV ratio p and the steady-state speed V_s as independent variables. The relative velocities of the traffic wave and the stationary road surface under different CAV ratios of the mixed traffic flow are calculated as shown in Figure 4.5.

It can be observed that the shapes of the CAV ratio curves for different steady-state speeds are similar, and the speed of the traffic wave increases with the steady-state speed of the traffic flow; under the same steady-state speed, the higher the CAV ratio is, the faster the backward transmission of the traffic wave is, that is, in a mixed traffic flow with a higher CAV ratio, the time for the traffic wave to affect the stability of the traffic flow is shorter, and the robustness of the traffic flow is higher.

4.5 SIMULATION TOOLS

Before autonomous vehicles (AVs) can be deployed, they need to undergo performance and safety testing in various scenarios. However, traditional road testing suffers from long testing cycles, high costs, and significant safety risks. To address these

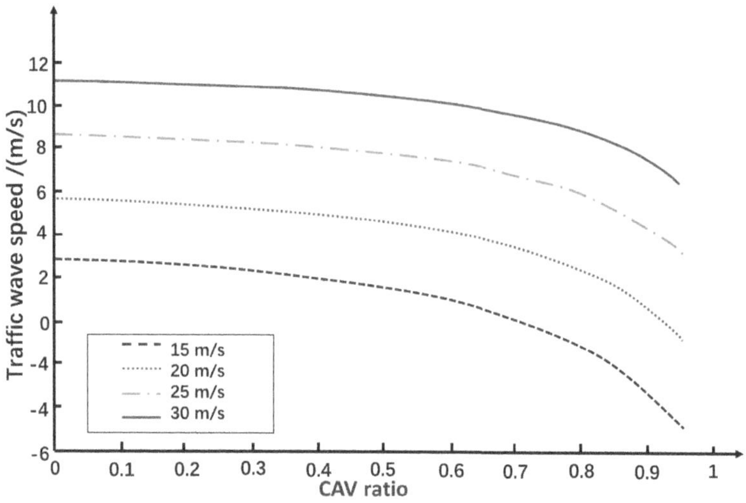

FIGURE 4.5 The velocity of traffic shock wave in mixed traffic flow.

shortcomings, simulation testing for autonomous driving has gradually become a key tool for designing, validating, and evaluating AV systems due to its safety, controllability, and repeatability. AV simulators are indispensable tools for conducting simulation tests for autonomous driving. These simulators are complex software toolkits used to simulate the driving and interaction behaviors of autonomous vehicles in different scenarios (Zhang et al., 2024). In scenarios involving mixed traffic flow, AV simulators are typically divided into three categories: vehicle dynamics simulation tools, vehicle sensor simulation tools, and traffic flow simulation tools.

4.5.1 Vehicle Dynamics Simulation Tools

Vehicle dynamics simulation tools focus on simulating and analyzing the dynamic behavior of vehicles in various driving scenarios. These tools are widely used in vehicle design, development, and testing stages to gain insights into the performance of vehicles in terms of handling, steering, acceleration, and other aspects under various conditions. Widely used AV vehicle dynamics simulators in academia and industry include CarMaker, CarSim, VTD, and Simulink, as shown in Table 4.2.

4.5.1.1 CarMaker

CarMaker (Ipg-automotive.com), developed by IPGAutomotive, is an advanced vehicle dynamics simulation software tool with a user-friendly interface. As illustrated in Figure 4.6, CarMaker is renowned for its intelligent simulation capabilities tailored for passenger vehicles. It encompasses precise vehicle models (engine, chassis, suspension, transmission, steering, etc.) and facilitates the generation of road infrastructure (IPGRoad), traffic environments (IPGTraffic), driver behaviors (IPGDriver), and vehicle maneuvers (IPGManeuver) through graphical user interface (GUI).

TABLE 4.2
Vehicle dynamics simulation tools

Name	Features
CarMaker	User-friendly interface; detailed Geographic Information System data application; support for the development and testing of advanced driver assistance systems.
CarSim	Used for automotive dynamics research and advanced driver assistance system (ADAS) development; accurate and stable vehicle dynamics modeling; support for various vehicle model configurations; integration with other simulation software; provides rich APIs.
VTD	Customizable 3D environment; rich sensor model library; open interface; support for HIL system integration; highly scalable.
Simulink	Support for HERE HD real-time map data and OpenDRIVE road network; simulation of various sensors; widely used in control algorithm development and decision-making; support for HIL and SIL testing.

FIGURE 4.6 CarMaker simulation interface.

Additionally, CarMaker can be used for the development and testing of advanced driver assistance systems, offering support for systems such as lane departure warning systems, night vision systems, and adaptive cruise control systems. It can also simulate complex traffic environment simulations, support various sensor types, and has the ability to execute sophisticated test scripts for challenging scenarios, replicating real-world road tests (Zhang & Guo, 2008; Feng et al., 2016).

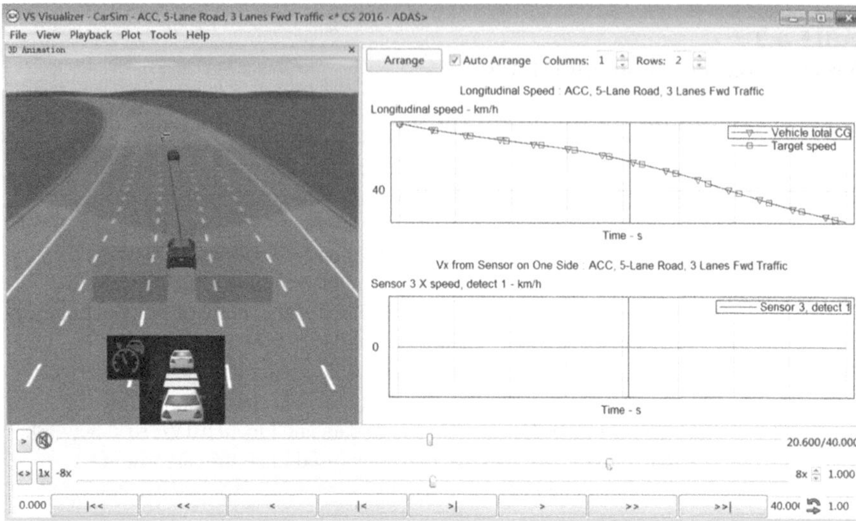

FIGURE 4.7 CarSim simulation interface.

4.5.1.2 CarSim

CarSim (www.carsim.com/) is widely used in automotive dynamics research, advanced driver assistance system (ADAS) development, and autonomous driving system verification. As depicted in Figure 4.7, CarSim excels in highly accurate and stable vehicle dynamics modeling, supporting various vehicle model configurations for passenger, commercial, and racing vehicles. Seamless integration with other simulation software such as MATLAB/Simulink and dSPACE, coupled with rich API provision, CarSim enables users to customize control algorithms and scenarios across multiple programming environments.

4.5.1.3 VTD

VTD (https://hexagon.com/products/virtual-test-drive), developed by VIRES in Germany, is a highly professional and powerful simulation platform tailored for the autonomous driving domain. As showcased in Figure 4.8, VTD provides customizable and realistic 3D environments, allowing the simulation of diverse traffic scenarios, weather conditions, and infrastructure elements. With an extensive sensor model library, open interfaces, and support for hardware-in-the-loop (HIL) system integration, VTD offers exceptional scalability and seamless integration with other autonomous driving simulation platforms, frameworks (e.g., Robot Operating System (ROS), MATLAB/Simulink), and vehicle dynamics simulators.

4.5.1.4 Simulink

MATLAB/Simulink Automated Driving Toolbox™ (www.mathworks.com/products/automated-driving.html) offers engineers and researchers a comprehensive set of tools for rapid design, simulation, and testing of ADAS and autonomous driving systems. Figure 4.9 illustrates Simulink's simulation interface, which supports the

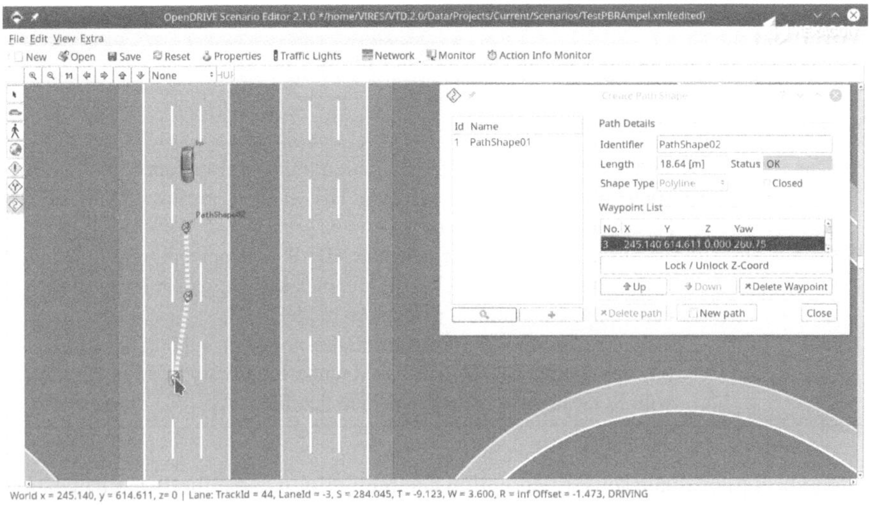

FIGURE 4.8 VTD simulation interface.

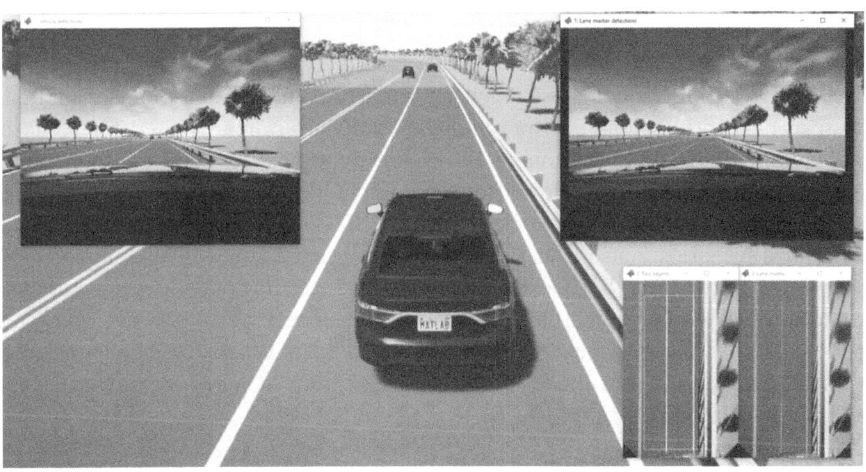

FIGURE 4.9 Simulink simulation interface.

import of HERE HD real-time map data and OpenDRIVE (Dupuis et al., 2010) road networks, enabling the creation of realistic 3D scenes and simulation of various sensors. Widely used in control algorithm development, sensor fusion, target tracking, path planning, decision-making, and supporting HIL and software-in-the-loop (SIL) testing, Simulink is a versatile solution in the autonomous driving domain.

4.5.2 Vehicle Sensor Simulation Tools

Vehicle sensor simulation tools are software applications used to simulate and test the performance of sensors in various scenarios. These tools allow engineers and

TABLE 4.3
Vehicle sensor simulation tools

Name	Features
PreScan	Detailed and accurate sensor simulation; multi-sensor support; compatibility with other software; HIL and VIL testing support; highly precise physical modeling.
dSPACE ASM	Support for autonomous driving and ADAS development; real-time model collection; precise modeling of complex scenes; sensor integration capability; environment perception model.
Gazebo	Widely used in robotics research and development; support for various physics engines and OGRE-based rendering library; tight integration with ROS; highly customizable and flexible.

researchers to evaluate sensor perception and recognition capabilities in virtual environments and validate sensor fusion and data processing algorithms. Commonly used vehicle sensor simulation tools include PreScan, dSPACE ASM, and Gazebo, as outlined in Table 4.3.

4.5.2.1 PreScan

PreScan (plm.sw.siemens.com/en-US/simcenter/autonomous-vehicle-solutions/prescan/) is a professional sensor simulation software primarily used for verifying and optimizing ADAS and autonomous driving algorithms. As shown in Figure 4.10, this software supports multiple sensor types, including LiDAR, radar, and cameras, and is compatible with MATLAB/Simulink, CarMaker, and other software. Users can customize scenarios, sensors, and road models as needed. PreScan also supports HIL and virtual-in-the-loop (VIL) testing, allowing engineers to combine actual hardware devices with simulated environments to more accurately assess the overall system performance. The software adopts highly accurate physical modeling, including vehicle dynamics, sensor models, and complex traffic scenarios and environments (Zhang & Guo, 2008).

4.5.2.2 dSPACE ASM

dSPACE ASM (www.dspace.com/en/pub/home.cfm) is designed to support the development and testing of autonomous driving and ADAS. As illustrated in Figure 4.11, this software provides finely tuned real-time models, including vehicle dynamics, environmental perception, traffic models, and sensor models. In the field of autonomous driving, dSPACE ASM stands out for its precise modeling of complex scenes and sensor integration capabilities. Its environment perception model can be used to generate realistic traffic scenarios, including various road types, weather conditions, traffic signs, and signals.

4.5.2.3 Gazebo

Gazebo, developed by Open Robotics, is an open-source simulation platform widely used in robotics research and development, including the field of

FIGURE 4.10 PreScan simulation interface.

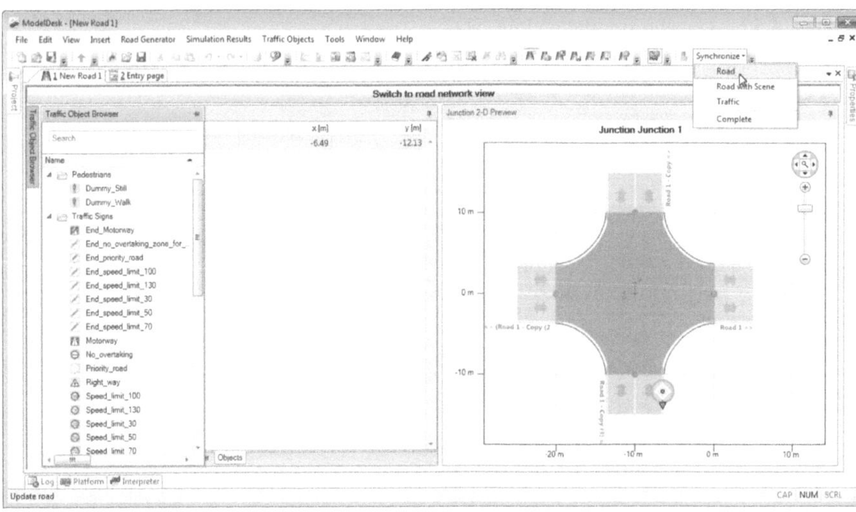

FIGURE 4.11 dSPACE ASM simulation interface.

autonomous driving (Koenig & Howard, 2004). The autonomous driving vehicle simulation platform built with Gazebo is depicted in Figure 4.12 (Streeting & Johnstone, 2010). This software supports various physics engines and the OGRE-based rendering library to ensure realistic physics simulation and visualization. With tight integration with ROS, developers can conveniently utilize ROS's various functionalities for simulation and control. Gazebo offers high customizability and flexibility, supporting the creation of complex 3D scenes, including various terrains, buildings, and roads.

FIGURE 4.12 Gazebo simulation interface.

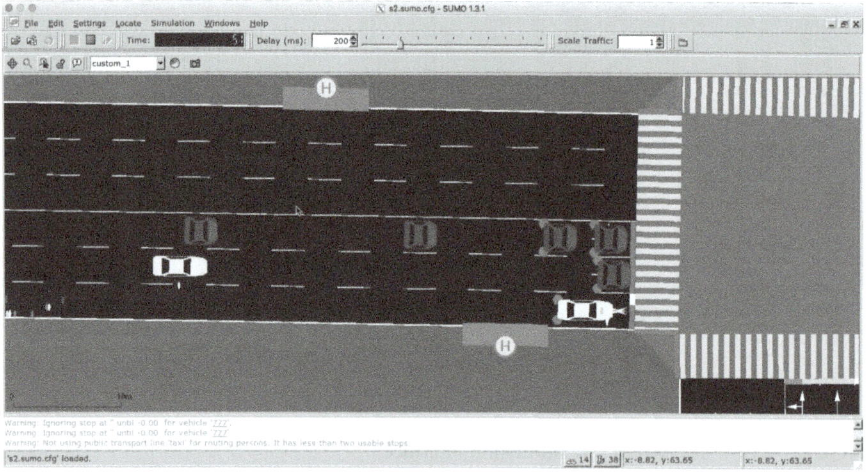

FIGURE 4.13 SUMO simulation interface.

4.5.3 Traffic Flow Simulation Tools

Traffic flow simulation tools are used to simulate and analyze the movement behavior of vehicles and pedestrians in traffic systems. In traditional traffic engineering, they are primarily used for intersection design and signal optimization, traffic planning and optimization, evaluation of traffic management strategies, and research on factors

TABLE 4.4
Traffic flow simulation tools

Name	Features
SUMO	Open-source microscopic urban traffic simulation software; simulates behavior of vehicles, pedestrians, and other entities; supports C++ development; integrates with various frameworks.
PTV VISSIM	Designed for complex traffic scenarios and multimodal traffic systems; simulates behavior of various traffic participants; supports V2I and V2V communication simulation; integrates with other software and hardware platforms.
MATSim	Open-source traffic simulation software for large-scale multi-agent transport systems; simulates scenarios with shared roads for autonomous and traditional vehicles; integrates with vehicle dynamics models, control algorithms, and HIL testing tools; predicts potential impacts of autonomous vehicles on congestion, road capacity, travel time, and environment.
Aimsun	Supports various driving scenarios; simulates complex road conditions and diverse behaviors of traffic participants; provides analysis of key performance indicators, fault modes, and effects.
TESS NG	Designed for Chinese drivers and traffic flow characteristics; high time update frequency; simulation of various traffic scenarios and tools; extensive development interface and customization features.

affecting traffic flow. In the field of autonomous driving, they are used for system testing, algorithm optimization and validation, interactive performance evaluation, and research on path planning and decision-making algorithms. Common traffic flow simulation tools include SUMO, PTV VISSIM, MATSim, Aimsun, and TESS NG, as summarized in Table 4.4.

4.5.3.1 SUMO

SUMO (Simulation of Urban Mobility, https://sumo.dlr.de/docs/index.html) is an open-source microscopic urban traffic simulation software used to simulate the behavior of vehicles, pedestrians, and other moving entities in urban traffic systems. Its simulation interface is shown in Figure 4.13. SUMO stands out for its microscopic, spatially continuous, temporally discrete characteristics and supports for C++ development (Hou et al., 2021). The software can integrate with other simulation software and frameworks, including vehicle dynamics simulators, the ROS (Quigley et al., 2009), and autonomous driving simulation platforms, enabling comprehensive analysis and evaluation of autonomous driving systems.

4.5.3.2 PTV VISSIM

PTV VISSIM (www.ptvgroup.com/en/products/ptv-vissim) is designed for complex traffic scenarios and multimodal traffic systems, as depicted in Figure 4.14. In the field of autonomous driving, PTV VISSIM accurately simulates the behavior

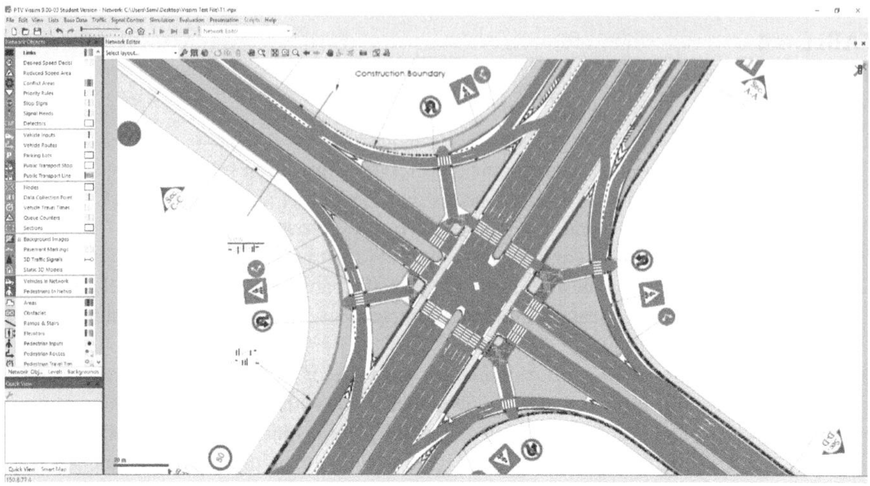

FIGURE 4.14 VISSIM simulation interface.

of various traffic participants through microscopic simulation techniques, providing researchers with realistic traffic environments to evaluate and optimize autonomous driving systems. The software supports the simulation of vehicle-to-infrastructure (V2I) and vehicle-to-vehicle (V2V) communication, facilitating the assessment of system effectiveness and safety. PTV VISSIM can integrate with other software and hardware platforms, enabling engineers and researchers to use VISSIM in conjunction with autonomous driving control systems, sensor simulation, HIL testing, and other tools for comprehensive system evaluation and optimization (Zhang & Guo, 2008).

4.5.3.3 MATSim

MATSim (Multi-Agent Transport Simulation, www.matsim.org/docs/) is a highly modular open-source traffic simulation software designed for large-scale multi-agent transport systems. Its software interface is depicted in Figure 4.15. MATSim simulates scenarios where autonomous and traditional vehicles share roads, allowing researchers to predict the potential impacts of autonomous vehicles on traffic congestion, road capacity, travel time, and the environment. Additionally, MATSim supports integration with vehicle dynamics models, vehicle control algorithms, and HIL testing tools, enabling multi-level assessment and optimization of autonomous driving systems.

4.5.3.4 Aimsun

Aimsun (CASAS et al., 2010; https://docs.aimsun.com/next/22.0.1/UsersManual/VehicleTypeEditing.html) supports various driving task scenarios, including urban traffic, highways, and rural roads, as shown in Figure 4.16, facilitating rapid iteration and performance improvement of autonomous driving systems. Aimsun can simulate complex road conditions and diverse behaviors of traffic participants, providing a

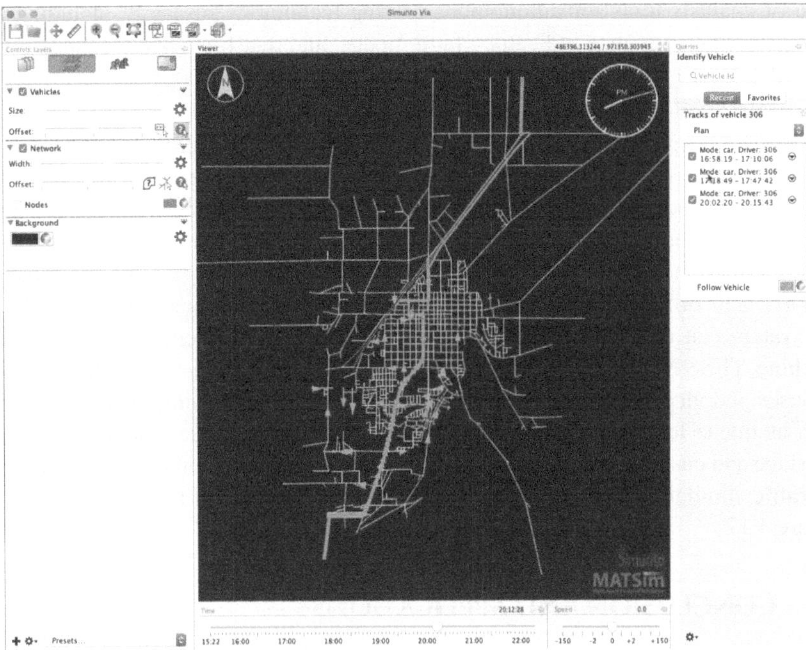

FIGURE 4.15 MATSim software interface.

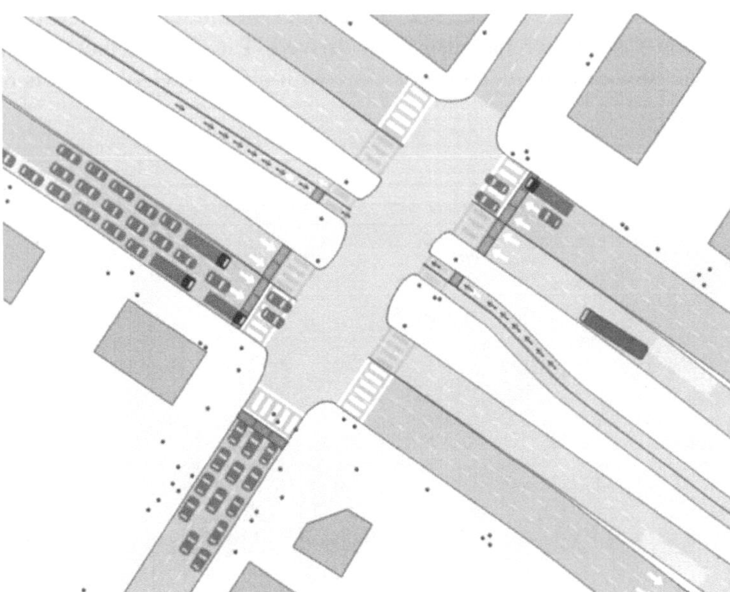

FIGURE 4.16 Aimsun simulation interface.

suite of analysis tools for autonomous driving technology research, including analysis of key performance indicators, fault modes, and real-time data visualization.

4.5.3.5 TESS NG

The TESS (Tongji Traffic Network Simulation System), a microscopic traffic simulation system, is a tool with completely independent intellectual property rights, designed specifically for Chinese drivers and traffic flow characteristics. Its features include convenient modeling, open interfaces, and user customization, enabling efficient simulation in various environments and supporting large-scale interchanges, airports, and other complex scenarios. With an update frequency of up to 20 Hz, the system can achieve real-time simulation of over 500 intersections on a single machine. TESS NG also supports the simulation of various transportation modes such as buses and airplanes, and their interactive behaviors, providing visual evaluations such as queue length and travel time. The system's robust secondary development interface and customization capabilities give it broad application potential in the field of traffic simulation (Song, 2020). The TESS NG simulation interface is shown in Figure 4.17.

4.6 CONCLUSION AND APPLICATIONS

In this chapter, the relationship between vehicle density and follow-through model variables for mixed traffic flows with different CAV ratios has been analyzed. The theoretical relationship between different following relationship ratios and CAV ratios was derived and verified. The basic diagram model of mixed traffic flow with different CAV ratios has also been derived. The enhancement of the maximum traffic flow rate in mixed traffic flow with different CAV ratios is obtained

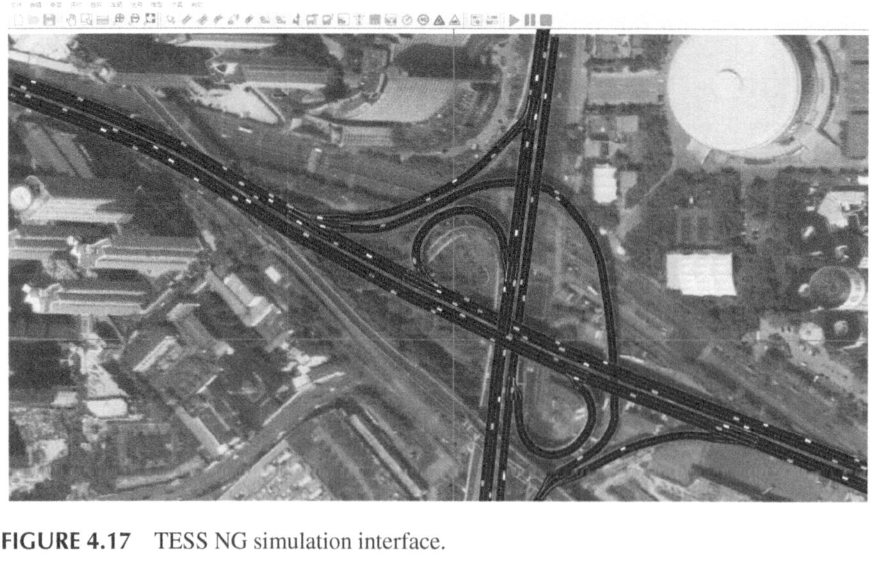

FIGURE 4.17 TESS NG simulation interface.

by the basic diagram model calculation. The LWR model has been introduced, and the traffic shock wave velocity calculation formula for the mixed traffic flow is derived.

At the same time, in the current development of traffic systems, there is a trend toward synergistic advancement through the combined use of vehicle dynamics simulation tools, vehicle sensor simulation tools, and traffic flow simulation tools. Vehicle dynamics simulation tools provide in-depth evaluations of vehicle performance by simulating vehicle motion under different road conditions. Meanwhile, vehicle sensor simulation tools replicate the perception capabilities of vehicle sensors under various environmental conditions, including lighting and weather variations. Combining these two with traffic flow simulation tools can create a comprehensive simulation environment capable of more accurately reproducing the operating scenarios of autonomous vehicles on real roads. This integrated approach not only drives continuous optimization of autonomous driving technology but also provides reliable tools for the intelligent development of traffic systems. This developmental trend of collaborative simulation is poised to provide the transportation industry with more comprehensive and precise testing and validation methods, accelerating the practical applications of autonomous driving and intelligent transportation. In the future, as these simulation tools further develop and integrate, they may make mixed traffic systems safer, smarter, and more efficient.

REFERENCES

Aimsun. (2023, April 6). User documentation. https://docs.aimsun.com/next/22.0.1/UsersManual/VehicleTypeEditing.html

An, G., & Talebpour, A. (2019, October). Lane-changing trajectory optimization to minimize traffic flow disturbance in a connected automated driving environment. In *2019 IEEE Intelligent Transportation Systems Conference (ITSC)* (pp. 1794–1799). IEEE.

Axhausen, K. W., Horni, A., & Nagel, K. (2016). *The Multi-Agent Transport Simulation MATSim* (p. 618). Ubiquity Press.

Basu, C., Yang, Q., Hungerman, D., et al. (2017) Do you want your autonomous car to drive like you? In *Proceedings of the 2017 ACM/IEEE International Conference on Human-Robot Interaction*, pp. 417–425.

CarMaker. (n.d.). Ipg-Automotive.Com. https://ipg-automotive.com/cn/products-solutions/software/carmaker/

Casas, J., Ferrer, J. L., Garcia, D., Perarnau, J., & Torday, A. (2010). Traffic simulation with aimsun. *Fundamentals of Traffic Simulation*, 173–232.

Chai, C., & Wong, Y. D. (2015). Fuzzy cellular automata model for signalized intersections. *Computer-Aided Civil and Infrastructure Engineering*, 30(12), 951–964.

Chen, H., & Wang, J. (2019). Automated highway driving lane-changing decision based on driver dissatisfaction. *Chinese Journal of Highway*, 32(12), 1–9.

dSPACE. (2023, April 6). dSPACE ASM. www.dspace.com/en/pub/home.cfm

Dupuis, M., Strobl, M., & Grezlikowski, H. (2010, August). Opendrive 2010 and beyond–status and future of the de facto standard for the description of road networks. In *Proceedings of the Driving Simulation Conference Europe* (pp. 231–242).

Ernst, D., Glavic, M., Capitanescu, F., & Wehenkel, L. (2008). Reinforcement learning versus model predictive control: A comparison on a power system problem. *IEEE Transactions on Systems, Man, and Cybernetics, Part B (Cybernetics)*, 39(2), 517–529.

Feng, G., Saisai, H., Xipeng, L., & University, C. (2016). Study on development technologies of driver assistance systems based on machine vision using NI EVS and PXI. *Foreign Electronic Measurement Technology*, 35(02), 30–34, 41.

Greenshields, B. D., Bibbins, J. R., Channing, W. S., & Miller, H. H. (1935, December). A study of traffic capacity. In *Highway Research Board Proceedings* (Vol. 14, No. 1, pp. 448–477).

Hou, D., Gao, L., Qian, Z., Li, Z., & Li, Y. (2021). Development and integration of vehicle model for heterogeneous traffic flow simulation platform based on SUMO. *Journal of Tongji University (Natural Science)*, 49(7), 957–963.

Huang, X., Sun, J., & Sun, J. (2018). A car-following model considering asymmetric driving behavior based on long short-term memory neural networks. *Transportation Research Part C: Emerging Technologies*, 95, 346–362.

Karnchanachari, N., Valls, M. I., Hoeller, D., & Hutter, M. (2020, July). Practical reinforcement learning for MPC: Learning from sparse objectives in under an hour on a real robot. In *Learning for Dynamics and Control* (pp. 211–224). PMLR.

Kerner, B. S. (1998). Experimental features of self-organization in traffic flow. *Physical Review Letters*, 81(17), 3797.

Kerner, B. S. (1999). Congested traffic flow: Observations and theory. *Transportation Research Record*, 1678(1), 160–167.

Kerner, B. S. (2000). Experimental features of the emergence of moving jams in free traffic flow. *Journal of Physics A: Mathematical and General*, 33(26), L221.

Kerner, B. S. (2004). *The Physics of Traffic: Empirical Freeway Pattern Features, Engineering Applications, and Theory*. Springer Science & Business Media.

Kerner, B. S., & Rehborn, H. (1996). Experimental properties of complexity in traffic flow. *Physical Review E*, 53(5), R4275.

Kerner, B. S., & Rehborn, H. (1997). Experimental properties of phase transitions in traffic flow. *Physical Review Letters*, 79(20), 4030.

Koenig, N., & Howard, A. (2004, September). Design and use paradigms for gazebo, an open-source multi-robot simulator. In *2004 IEEE/RSJ International Conference on Intelligent Robots and Systems (IROS)* (IEEE Cat. No. 04CH37566) (Vol. 3, pp. 2149–2154). IEEE.

Kuefler, A., Morton, J., Wheeler, T., & Kochenderfer, M. (2017, June). Imitating driver behavior with generative adversarial networks. In *2017 IEEE Intelligent Vehicles Symposium (IV)* (pp. 204–211). IEEE.

Li, T., Wu, J., Chan, C. Y., Liu, M., Zhu, C., Lu, W., & Hu, K. (2020). A cooperative lane change model for connected and automated vehicles. *IEEE Access*, 8, 54940–54951.

Lighthill, M. J., & Whitham, G. B. (1955). *On kinematic waves II: A theory of traffic flow on long crowded roads. Proceedings of the Royal Society A*, 229, 317–345.

Lin, Y., McPhee, J., & Azad, N. L. (2020). Comparison of deep reinforcement learning and model predictive control for adaptive cruise control. *IEEE Transactions on Intelligent Vehicles*, 6(2), 221–231.

Liu, H., Kan, X. D., Shladover, S. E., Lu, X. Y., & Ferlis, R. E. (2018). Modeling impacts of cooperative adaptive cruise control on mixed traffic flow in multi-lane freeway facilities. *Transportation Research Part C: Emerging Technologies*, 95, 261–279.

Liu, P., Kurt, A., & Ozguner, U. (2018). Distributed model predictive control for cooperative and flexible vehicle platooning. *IEEE Transactions on Control Systems Technology*, 27(3), 1115–1128.

Liu, Y., Guo, J., Taplin, J., & Wang, Y. (2017). Characteristic analysis of mixed traffic flow of regular and autonomous vehicles using cellular automata. *Journal of Advanced Transportation*, 2017, Article 8142074.

Lu, H., Wang, C., & Lu, Y. (2014). *Mechanisms and Countermeasures of Urban Traffic Congestion*. China Architecture & Building Press.

Luo, F., Larson, J., & Munson, T. (2018). Coordinated platooning with multiple speeds. *Transportation Research Part C: Emerging Technologies*, 90, 213–225.

Luo, L. H., Liu, H., Li, P., & Wang, H. (2010). Model predictive control for adaptive cruise control with multi-objectives: comfort, fuel-economy, safety and car-following. *Journal of Zhejiang University Science A*, 11(3), 191–201.

Luo, Y., Yang, G., Xu, M., Qin, Z., & Li, K. (2019). Cooperative lane-change maneuver for multiple automated vehicles on a highway. *Automotive Innovation*, 2, 157–168.

Ma, L., & Qu, S. (2020). A sequence to sequence learning based car-following model for multi-step predictions considering reaction delay. *Transportation Research Part C: Emerging Technologies*, 120, 102785.

Mathworks. (2023, April 6). Automated driving toolbox. www.mathworks.com/products/automated-driving.html

MATSim. (2023, April 6). User documentation. www.matsim.org/docs/

Mechanical Simulation. (2023, April 6). CarSim. www.carsim.com/

Ni, J., Han, J., & Dong, F. (2020). Multivehicle cooperative lane change control strategy for intelligent connected vehicle. *Journal of Advanced Transportation*, 2020, 1–10.

Pan, F., & Bao, H. (2019, December). Reinforcement learning model with a reward function based on human driving characteristics. In *2019 15th International Conference on Computational Intelligence and Security (CIS)* (pp. 225–229). IEEE.

Pipes, L. A. (1953). An operational analysis of traffic dynamics. *Journal of Applied Physics*, 24(3), 274–281.

PTV VISSIM. (n.d.). Ptvgroup.Com. www.ptvgroup.com/en/products/ptv-vissim

Quigley, M., Conley, K., Gerkey, B., Faust, J., Foote, T., Leibs, J., ... & Ng, A. Y. (2009, May). ROS: An open-source Robot Operating System. In *ICRA Workshop on Open Source Software* (Vol. 3, No. 3.2, p. 5).

Richards, P. I. (1956). Shock waves on the highway. *Operations Research*, 4(1), 42–51.

Schönhof, M., & Helbing, D. (2007). Empirical features of congested traffic states and their implications for traffic modeling. *Transportation Science*, 41(2), 135–166.

Simcenter Prescan software simulation platform. (n.d.). Siemens Digital Industries Software. https://plm.sw.siemens.com/en-US/simcenter/autonomous-vehicle-solutions/prescan/

Song, Z. (2020). *Adaptability Analysis on Micro-Simulation of High-Density Traffic Flow – A Comparative Study of Three Micro-Simulation Packages* [Doctoral Dissertation]. Shanghai: Tongji University.

Streeting, S., & Johnstone, B. (2010). OGRE – open source 3D graphics engine. *OGRE*, 26, 09–12. www.ogre3d.org/

SUMO Documentation. (n.d.). Dlr.De. https://sumo.dlr.de/docs/index.html

Sun, K., Zhao, X., & Wu, X. (2021). A cooperative lane change model for connected and autonomous vehicles on two lanes highway by considering the traffic efficiency on both lanes. *Transportation Research Interdisciplinary Perspectives*, 9, 100310.

Takahama, T., & Akasaka, D. (2018). Model predictive control approach to design practical adaptive cruise control for traffic jam. *International Journal of Automotive Engineering*, 9(3), 99–104.

Tanveer, M., Kashmiri, F. A., Naeem, H., Yan, H., Qi, X., Rizvi, S. M. A., ... & Lu, H. (2020). An assessment of age and gender characteristics of mixed traffic with autonomous and manual vehicles: A cellular automata approach. *Sustainability*, 12(7), 2922.

VTD. (n.d.). Virtual test drive. https://hexagon.com/products/virtual-test-drive

Wang, D. H., & Jin, S. (2012). Review and outlook of modeling of car following behavior. *Zhongguo Gonglu Xuebao(China Journal of Highway and Transport)*, 25(1), 115–127.

Wang, D. H., Tao, P. F., Jin, S., & Ma, D. F. (2011). Method of calibrating and validating car-following model. *Journal of Jilin University (Engineering and Technology Edition)*, 1, 59–65.

Wang, X., Sun, P., & Zhang, X. (2020). Parameter calibration of highway following model based on natural driving data. *Chinese Journal of Highway*, 33(05), 132–142.

Wegener, M., Koch, L., Eisenbarth, M., & Andert, J. (2021). Automated eco-driving in urban scenarios using deep reinforcement learning. *Transportation Research Part C: Emerging Technologies*, 126, 102967.

Yang, D., Zhu, L. L., Yu, D., Yang, F., & Pu, Y. (2014). An enhanced safe distance car-following model. *Journal of Shanghai Jiaotong University (Science)*, 19, 115–122.

Yang, Q. I., & Koutsopoulos, H. N. (1996). A microscopic traffic simulator for evaluation of dynamic traffic management systems. *Transportation Research Part C: Emerging Technologies*, 4(3), 113–129.

Zhang, J., & Guo, F. (2008). *Research on ABS Control Algorithm Based on Model Reference Adaptive PID Control. Chinese Academy of Automotive Engineering Annual Meeting of 2008*. Beijing: Chinese Academy of Automotive Engineering (pp. 1–3) (in Chinese).

Zhang, K., Chang, C., & Wang, S. (2024). Review of autonomous vehicle simulators: Capabilities, challenges, and development directions. *Journal of Transportation Engineering and Information*, 1–30. https://doi.org/10.19961/j.cnki.1672-4747.2023.04.007

Zhou, J., & Zhu, F. (2021). Analytical analysis of the effect of maximum platoon size of connected and automated vehicles. *Transportation Research Part C: Emerging Technologies*, 122, 102882.

Zhou, M., Qu, X., & Jin, S. (2016). On the impact of cooperative autonomous vehicles in improving freeway merging: A modified intelligent driver model-based approach. *IEEE Transactions on Intelligent Transportation Systems*, 18(6), 1422–1428.

Zhou, M., Qu, X., & Li, X. (2017). A recurrent neural network based microscopic car following model to predict traffic oscillation. *Transportation Research Part C: Emerging Technologies*, 84, 245–264.

Zhou, M., Yu, Y., & Qu, X. (2019). Development of an efficient driving strategy for connected and automated vehicles at signalized intersections: A reinforcement learning approach. *IEEE Transactions on Intelligent Transportation Systems*, 21(1), 433–443.

Zhou, Y. J., Zhu, H. B., Guo, M. M., & Zhou, J. L. (2020). Impact of CACC vehicles' cooperative driving strategy on mixed four-lane highway traffic flow. *Physica A: Statistical Mechanics and Its Applications*, 540, 122721.

Zhu, B., Jiang, Y., Zhao, J., Chen, H., & Deng, W. W. (2019). A car-following control algorithm based on deed reinforcement learning. *China Journal of Highway and Transport*, 32(006), 53–60.

Zhu, M., Wang, Y., Pu, Z., Hu, J., Wang, X., & Ke, R. (2020). Safe, efficient, and comfortable velocity control based on reinforcement learning for autonomous driving. *Transportation Research Part C: Emerging Technologies*, 117, 102662.

Zou, Z., & Yang, D. (2002). Lane transformation model in microscopic traffic simulation. *China Highway Journal*, 2, 108–111.

5 Safety of Mixed Traffic

Evidence from Empirical Studies

Chunxi Huang, Tianhao Li, and Dengbo He

5.1 SIMULATION-BASED MIXED TRAFFIC SAFETY RESEARCH

The simulation-based approach has been intensively used for mixed traffic research in the past few years and has shown a great advantage in exploring the effects of connected and automated vehicles (CAVs) on mixed traffic safety, especially under different penetration rates of CAVs. For example, by using different car-following models to capture characteristics of different types of vehicles (e.g., human-driven vehicles (HDVs) and cooperative adaptive cruise control (CACC) vehicles), Yao et al. (2020) investigated the safety of mixed traffic flow on expressways under different penetration rates of CAVs using multiple safety measures (i.e., standard deviation of vehicle speed, time-exposed rear-end crash risk, time-exposed time-to-collision (TTC), and time-integrated TTC), and they found that the increase of CACC vehicles reduces traffic safety risks significantly when the penetration rate exceeds 50%. Similarly, Ye and Yamamoto (2019) analyzed the impact of CAVs on traffic safety under various penetration rates, in which the frequency of dangerous situations and TTC were used as indicators of traffic safety. They revealed that the condition of traffic safety is greatly improved with the increase in CAV penetration rate, while a more cautious car-following strategy of the CAV would contribute more to traffic safety. Jiang et al. (2023) explored the impact of platoon size on the safety of mixed traffic flows through numerical simulations; and they found that the effect of platoon size of CAVs on the safety of mixed traffic flow was influenced by the CAV penetration rate. Specifically, when the platoon size reaches 2–3, the mixed traffic flow is safer when the penetration rate of CAVs is lower than 40%. By using the CACC car-following models, Arvin et al. (2021) also found that the safety of mixed traffic flow improved substantially with a penetration rate of more than 40%. Garg and Bouroche (2023) investigated the detrimental effects of unreliable vehicle-to-vehicle (V2V) communication and large reaction times on the safety of mixed traffic flow evaluated by the number of safety conflicts, and they found that, at 70% penetration rates, CAVs could reduce 66.67% of the safety conflicts in the presence of unreliable V2V communication (i.e., packet drop of 70%) and large reaction times (i.e., 0.5 second) by adopting a very cautious car-following strategy.

However, there are several major limitations of simulation-based studies. First, the behaviors of HDVs and CAVs in the simulation were typically modeled using various

DOI: 10.1201/9781032674667-5

car-following models (e.g., Intelligent Driver Model, Gipps Model, Wiedemann Model, CACC Model), while these models posed different assumptions and may not fully reflect the behaviors of vehicles under real conditions. Second, human drivers may have different attitudes toward CAVs and may interact with CAVs differently, and there might even be potential adaptation among human drivers, which is difficult for simulation-based studies to account for. Next, most simulation-based studies assumed that CAVs operated in traffic scenarios with perfect communication or with only a very small number of vehicles having imperfect communication, which may not be the case in real conditions. Finally, existing studies using simulation-based methods intensively considered the effects of penetration rates on the safety of mixed traffic flow, while it may take a long time to reach the penetration rates settled in those studies.

Therefore, to provide more realistic insights about the current mixed traffic safety, in this chapter, we will mainly review the research that adopted empirical methods, that is, a method that gains knowledge or obtains empirical evidence through direct and indirect observation or experience. Specifically, we will discuss mixed traffic safety from two perspectives: Mixed traffic safety analysis (i.e., crash analysis and surrogate measures analysis in mixed traffic) and mixed traffic behavioral analysis. We will discuss the major findings from existing research that focused on safety in mixed traffic.

5.2 REACTIVE SAFETY ANALYSIS OF MIXED TRAFFIC

As many companies (e.g., Waymo in the US and Baidu in China) have been testing their automated vehicles (AVs) on publicly open roads, many AV-involved crashes (or mixed traffic crashes) have already happened in the past few years. There were several public AV-involved crash report datasets available and interested readers could refer to Table 2.3 in this book for more information. We first discussed the approaches that existing studies used to analyze AV-involved crashes and highlighted some useful considerations for crash analysis. Furthermore, we summarized the characteristics of mixed traffic crashes from the following four perspectives: prevalence of mixed traffic crashes, dominant role of rear-end crashes, factors associated with mixed traffic crashes, and factors affecting mixed traffic crash outcomes.

5.2.1 Approaches for Crash Analysis

Previous studies have adopted various approaches to analyze AV crashes. In the early stages of AV testing (e.g., 2014–2018), AV crashes were mainly analyzed through descriptive statistics (Dixit et al., 2016; Favarò et al., 2017), potentially due to the limited sample size at that time. With the increasing number of crash records, some studies tried to identify the contributing factors of crash severity and collision type in AV crashes through statistical modeling or machine learning methods, such as logistic regression (Boggs et al., 2020; Wang and Li, 2019; Xu et al., 2019), classification and regression tree (CART) (Chen et al., 2020; Wang and Li, 2019), cluster analysis (Song et al., 2021), decision tree (Ashraf et al., 2021), association rule method (Ashraf et al., 2021; Liu et al., 2024), and extreme gradient boosting (XGBoost) (Chen et al., 2020).

TABLE 5.1
Summary of approaches used by previous studies to analyze mixed traffic crashes

Type of approaches	Approaches and corresponding studies
1. Descriptive statistics	Descriptive statistics (Dixit et al., 2016; Favarò et al., 2017)
2. Statistical modelling	Logistic regression (Boggs et al., 2020; Huang et al., 2023b; Wang and Li, 2019; Xu et al., 2019)
3. Machine learning methods	Classification and regression tree (CART) (Chen et al., 2020; Wang and Li, 2019); Cluster analysis (Song et al., 2021); Decision tree (Ashraf et al., 2021); Association rule method (Ashraf et al., 2021; Liu et al., 2024); Extreme gradient boosting (XGBoost) (Chen et al., 2020)
4. Text-analytic-based methods	Topic modelling and text mining (Alambeigi et al., 2020; Boggs et al., 2020; Kutela et al., 2022b; Lee et al., 2023)
5. Bayesian-related methods	Hierarchical Bayes logit models (Boggs et al., 2020); Bayesian latent class models (Das et al., 2020); Bayesian networks (Kutela et al., 2022a; Lee et al., 2023)

There were also some studies trying to investigate the characteristics of AV crashes by performing text-analytic-based methods on the narratives of AV crash reports (Alambeigi et al., 2020; Boggs et al., 2020; Kutela et al., 2022b; Lee et al., 2023). Some studies also adopted Bayesian-related methods to serve different purposes. For example, the hierarchical Bayes logit models were used for accounting for unobserved heterogeneity (Boggs et al., 2020), Bayesian latent class models were used to identify crash patterns (Das et al., 2020), and Bayesian networks (BN) to reveal interrelationships among influential factors of AV crashes (Kutela et al., 2022a; Lee et al., 2023). Table 5.1 summarizes the approaches used by previous studies to analyze mixed traffic crashes.

5.2.2 Characteristics of Mixed Traffic Crashes

5.2.2.1 Prevalence of Mixed Traffic Crashes

As the most representative and intuitive indicator of mixed traffic safety, the prevalence of crashes was widely evaluated using the crash rate data. Schoettle and Sivak (2015) conducted the first preliminary analysis to investigate the crash rates of AVs operated by three companies (i.e., Google, Delphi, and Audi) in California, and they further compared it to the crash rate of conventional vehicles in the US in 2013. They found that AVs have a higher crash rate (i.e., 9.1 crashes per million miles) than conventional vehicles (i.e., 4.1 crashes per million miles), though not statistically significant. It should be noted that, at the time, only 11 AV-involved crashes over 1.2 million miles were reported. In other words, the accumulated mileage of AVs was limited, and their exposure may not be comparable to the exposure of conventional vehicles.

Later, in 2016, Blanco et al. (2016) conducted another more comprehensive comparison between crash rates of AVs and conventional vehicles. They used internal data (including written reports, video, and vehicle kinematic data) from Google's Self-Driving Car program (about 1.3 million miles) and the data from the Second Highway Research Program (SHRP 2) Naturalistic Driving Study (NDS) (Blanco et al., 2016). Initially, they found that the AV crash rate (i.e., 8.7 crashes per million miles) was higher than that of conventional vehicles (i.e., 1.9 crashes per million miles) when the crash severity and reporting requirements were not controlled. However, when unreported crashes were controlled, the AV crash rates (i.e., 3.2 crashes per million miles) were lower than that of conventional vehicles (i.e., 4.2 crashes per million miles). According to the classifications of crash severity in SHRP 2 NDS, they also compared the crash severity-controlled AV crash rates to the SHRP 2 age-adjusted crash rates. Results indicated that AVs had lower rates of severe crashes (i.e., 1.6 crashes per million miles for both Level 1 and Level 2 crashes) when compared to conventional vehicles (i.e., 2.5 crashes per million miles for Level 1 crashes and 3.3 crashes per million miles for Level 2 crashes). However, the rates of less severe AV crashes (i.e., Level 3 crashes, 5.6 crashes per million miles) were significantly lower than the rates of conventional vehicles (i.e., 14.4 crashes per million miles).

Based on the crash reports collected by the California Department of Motor Vehicles (CA-DMV) between September 2014 and November 2015 (in total of 12 crashes), Dixit et al. (2016) estimated that the Google AV crash rate was 1 crash for every 47,148 miles, which was much higher than the crash rate of conventional vehicles (i.e., 1 crash for every 2.07 million miles) estimated from the California Highway Patrol (CHP) Safety Database from 2014 (Dixit et al., 2016). Two potential issues may influence the results: (1) the crash severity was not adjusted, as CA-DMV AV crash reports include all crashes while CHP crash reports only include crashes that met the reporting criteria; (2) the AVs were tested in the local streets in CA-DMV AV crash reports while the CHP database only covered State, US, and Interstate roads.

Favarò et al. (2017) also estimated the AV crash rates using the crash reports collected by CA-DMV between September 2014 and March 2017 (in total 26 crashes), and they found that the AV crash rates were 42,017 miles per crash (Favarò et al., 2017). To conduct the comparison, they estimated the crash rates of conventional vehicles based on the data from the National Highway Traffic Safety Administration (NHTSA) and Federal Highway Administration (FHWA), and they found that the crash rates of conventional vehicles (i.e., 500,000 miles per crash) were lower than that of AVs. Similarly, Banerjee et al. (2018) compared the AV crash rates between 2014 and 2017 with the crash rates of conventional vehicles from NHTSA and FHWA (Banerjee et al., 2018), they found that the Google AV crash rates (i.e., 41.4 crashes per million miles) were 20 times higher than the crash rates of conventional vehicles (i.e., 2 crashes per million miles). However, both Favarò et al. (2017) and Banerjee et al. (2018) compared all AV-involved crashes with police-reported crashes of conventional vehicles while not adjusting for the underreporting issue of conventional vehicle crashes.

To account for the under-reporting issue of conventional vehicle crashes, Teoh and Kidd (2017) coded Google AV crashes by type and severity through narratives released by Google, and they calculated the crash rates using crashes that were severe enough to be reportable to police. These police-reportable AV crash rates were compared with policed-reported crash rates for conventional vehicles between 2009 and 2015 in several regions, which included Mountain View, Santa Clara County, California, and the

US. They found that AV crash rates (i.e., 2.19 crashes per million miles) were lower than the crash rates of conventional vehicles in the US (i.e., 3.59 crashes per million miles) and Mountain View (i.e., 6.06 crashes per million miles), while higher than the crash rates of conventional vehicles in Santa Clara County (i.e., 1.35 crashes per million miles) and California (i.e., 1.92 crashes per million miles). The authors also compared the Google AV crash rates with the crash rates of conventional vehicles in SHRP 2 NDS, and results indicated that the Google AV crash rates (i.e., 9.65 crashes per million miles) were significantly lower than the crash rates of conventional vehicles in SHRP 2 NDS (i.e., 27.60 crashes per million miles). However, it should be noted that the crash rates of conventional vehicles in SHRP 2 NDS in this study are not age-adjusted, thus might be overcounting as the SHRP 2 NDS oversampled teen and elderly drivers who were known to have higher crash risks (Regev et al., 2018).

Given the high prevalence of rear-end collisions among AV-involved crashes, Goodall (2021) specifically compared the AV struck-from-behind crash rates with the national crash rates of conventional vehicles using the same crash definitions. The AV crash rates were filtered and extracted from the CA-DMV dataset, while the crash rates of conventional vehicles were calculated based on the SHRP 2 NDS, after weighing the age by the US driving population. The AVs were found to be struck from behind at 4.8 times the rate of conventional vehicles, while the AV crash rates were 5 times higher for urban driving when controlling for the driving environment. One potential reason for this could be that AVs were tested more in urban areas. In another study, Goodall (2024) shifted the focus from AVs to partial AVs (i.e., vehicles equipped with advanced driver assistance systems) (Goodall, 2024), in which they compared the crash rates of vehicles equipped with Society of Automobile Engineers (SAE) Level 1 automation and SAE Level 2 automation, with road type and age controlled. It was found that Level 2 vehicles only decreased the crash rates by 10% when controlling for different rates of freeway driving, even though the Level 2 vehicles were claimed to have a 43% lower crash rate than Level 1 vehicles (Goodall, 2024).

Table 5.2 summarizes the comparison of crash rates of AVs and conventional vehicles in existing studies. It can be observed that the results were somewhat mixed as different studies used different data sources for calculating the crash rates of AVs and conventional vehicles, and some studies adopted the adjustment to account for underreporting (e.g., Teoh and Kidd, 2017) and imbalanced age distribution (e.g., Goodall, 2021) while some studies did not. Therefore, it is difficult to judge the safety performance of AVs based on the current studies, as technologies continue to mature. Inspired by existing studies, we highly encourage future studies to account for potential bias while estimating crash rates (i.e., considering the exposure and underreporting crashes) and make fair comparisons.

At the same time, as a macroscopic indicator, the crash rate alone can provide the overall map of mixed traffic safety, while it may not reveal factors leading to crashes from a microscopic perspective. To guide the design and optimization of countermeasures that can improve the safety of mixed traffic, researchers tried to investigate other characteristics of mixed traffic crashes. We summarized the detailed characteristics of mixed traffic crashes in the following three perspectives: the dominant role of rear-end crashes, factors associated with AV crashes, and factors associated with AV crash outcomes.

TABLE 5.2
Comparison of crash rates of AVs and conventional vehicles

AV crashes source	Crash rates of AVs in per million miles [95% confidence interval]	Conventional vehicle crashes source	Crash rates of conventional vehicles in per million miles [95% confidence interval]	Source
All crashes	9.1 [4.5, 16.3]	Police-reportable crashes	4.1 [4.5, 16.3]	Schoettle and Sivak (2015)
Severity-adjusted Waymo crashes	3.2 [0.4, 11.4]	NHTSA Police-reportable crashes	4.2 [2.8, 9.9]	Blanco et al. (2016)
Severity-adjusted Waymo crashes	3.2 [0.4, 11.4]	Age-adjusted SHRP 2 Police-reportable crashes	5.8 [4.7, 7.0]	Blanco et al. (2016)
All crashes	21.21	CHP reported crashes with a limited range	0.48	Dixit et al. (2016)
All crashes	23.80	NHTSA and FWHA reported crashes	2.0	Favarò et al. (2017)
All crashes	41.4	NHTSA and FWHA reported crashes	2.0	Banerjee et al. (2018)
All crashes	9.65 [5.81, 15.07]	All SHRP 2 crashes (not adjusted)	27.60 [25.77, 29.53]	Teoh and Kidd (2017)
Police-reportable crashes	2.19 [0.44, 6.39]	Police-reportable crashes in US	3.59 [2.31, 4.87]	Teoh and Kidd (2017)
All Google rear-end struck	7.11 [3.89, 11.93]	SHRP 2 rear-end struck	2.65 [2.10, 3.29]	Teoh and Kidd (2017)
All AV struck-from-behind crashes	17.2 [14.2, 20.7]	Age-adjusted SHRP 2 struck-from-behind crashes	3.6 [3.0, 4.3]	Goodall (2021)

5.2.2.2 The Dominant Role of Rear-End Crashes

Rear-end crashes dominated the AV-involved mixed traffic crashes: One of the most representative characteristics of the mixed traffic crashes was that rear-end crashes dominated the AV-involved mixed traffic crashes. For example, Lee et al. (2023) found that 51.9% of 260 AV crashes from 2019 to 2021 were rear-end crashes, and when there was a crash, AVs were more vulnerable to rear-end collisions in the automated driving mode than in the conventional mode (Lee et al., 2023); Kutela et al. (2022a) found that 63.5% of 333 AV crashes from 2017 to 2021 were rear-end crashes, while AVs were at fault for only 14.4% of the crashes. Xu et al. (2019) found that two predominant collision types among AV crashes were rear-end and sideswipe crashes, which accounted for 57.5% and 28.8% of AV crashes, respectively (Xu et al., 2019). By comparing AV crashes with conventional vehicle crashes, Liu et al. (2021) found that the most frequent AV pre-crash scenarios were rear-ended collisions (52.46%) and lane-change collisions (18.85%), with the proportion of AVs rear-ended by conventional vehicles occurring with a frequency 1.6 times that of conventional vehicles (Liu et al., 2021). Similar findings were also obtained in Favarò et al. (2017) and Huang et al. (2024).

Situations when AVs being rear-ended by HVs: Given the high prevalence of rear-end crashes, some studies tried to understand the characteristics of these rear-end crashes. For instance, based on the AV crash reports from 2015 to 2020 collected by CA-DMV, McCarthy (2022) found that most AV crashes are the fault of human drivers, with AVs being rear-ended by HVs, which implies that human drivers appear less able to anticipate the AV's more conservative driving style. More specifically, several studies also revealed that most rear-end crashes occurred when AVs were stopped at intersections, while HVs were proceeding straight through the intersection (Ashraf et al., 2021; Liu et al., 2024). Based on a crash sequence analysis of 168 AV crashes reported to CA-DMV from 2015 to 2019, Song et al. (2021) found that the most representative pattern in AV crashes was "collision following AV stop". Through analyzing the AV crash reports collected by NHTSA between July 2021 and May 2022, Huang et al. (2023) also found that the likelihood of an AV-involved crash being a rear-end collision would be the highest when the following HV was proceeding straight. With 251 AV crash reports from CA-DMV as of February 2020, Ashraf et al. (2021) extracted the pre-crash motion of AV crashes using the decision tree and association rule methods, by which they found that the most frequent types of AV crashes were rear-end crashes and predominantly occurred at intersections when AVs were stopped and engaged in the automated driving mode. Compared to conventional vehicles, AVs were less likely to be involved in a collision from the front of the vehicle, such as hitting an object ahead (Novat et al., 2023). This important feature of AVs being struck from behind in AV-involved crashes was also evidenced by the safety performance data of Waymo (Schwall et al., 2020).

Various factors were found to be associated with AV-involved rear-end crashes: Several studies further explored the factors that were associated with rear-end crashes. The most common phenomenon was AV rear-ended by HVs. For example, Lee et al. (2023) investigated the interrelationships among pre-crash conditions, and they found that AVs were more likely to be rear-ended in the automated driving mode compared to the conventional mode, which means that AVs might not be capable enough to avoid rear-end collisions compared to HVs. At the same time, Lee et al. (2023) also revealed that there was no significant difference in the chance of rear-end collisions when comparing the conventional mode and manual disengagement, which means that an AV will be less likely to have a rear-end collision when it hands over the control to a human driver in case of emergency. Similarly, through analysis of event transition in the crash sequences, Song et al. (2021) found that the disengagement had a transition probability of 68% to an immediate collision, which has also been observed in Boggs et al. (2020). Specifically, Boggs et al. (2020) found that when the automated driving system was engaged and remained engaged, the likelihood of an AV-involved rear-end crash was substantially higher compared to a conventionally driven AV or when the driver disengages the automated driving system before a crash. Xu et al. (2019) also found that AV crashes that occurred under the automated driving mode were more likely to be rear-end crashes, and the crashes were more likely to be rear-end crashes when AVs were stopped or when AVs had turning movements (Xu et al., 2019). Besides, by comparing AV crashes with conventional vehicle crashes, Novat et al., (2023) revealed that AVs were more likely to be involved in rear-end collisions in signalized intersections and lanes with less than 45 miles per hour (mph) speed limit. Petrović et al. (2020) also indicated that drivers' typical errors of conventional vehicles in AV crashes were "unsafe speed" and "following too closely". Kutela et al. (2022a) also found that the chances of rear-end crashes were lower in the parking lot, and when nonvehicular participants were involved, but increased in high-traffic flow roadways.

Based on the findings from existing studies, it is straightforward to see the dominant role of rear-end crashes in AV crashes, and in most cases, the AVs were rear-ended by HVs at intersections when AVs stopped or drove slowly. Given the factors associated with AV-involved rear-end crashes, the reasons behind this phenomenon were twofold: (1) from the perspective of AV, current AVs had limited perception and prediction capability, thus adopted comparably conservative driving styles, and usually would stop or drive slowly at the intersection faced with the complex environment and interaction with various road agents (e.g., pedestrians, HVs, traffic lights); (2) from the perspective of HV, human drivers may not develop an appropriate mental model of current AVs (Huang et al., 2023), thus might drive in a more aggressive way (e.g., following too close with high speed) and could not anticipate AVs' behaviors.

5.2.2.3 Factors Associated with AV Crashes

To provide more insights into the design of relevant countermeasures, some studies also investigated the factors associated with AV crashes, and they found that various environment-related and motion-related factors were associated with AV crashes.

In terms of environmental-related factors, Kutela et al. (2022b) found that crosswalks, intersections, and traffic signals are the key predictors of vulnerable road users (VRUs)-AV-related crashes, in which bicyclists and scooterists were more likely to be involved in the AV crashes directly, while pedestrians were more likely to be involved in the AV crashes indirectly. In another study, Kutela et al. (2022a) shifted the research focus to vehicle at-fault and collision type of AV crashes, and they found that the chance of an AV being at fault seems higher in parking locations, and during driving on wet pavements in unclear weather conditions. Liu et al. (2024) also found that AV crashes are more likely to occur in poor lighting conditions (e.g., night), unsignalized intersections, and two-way roads without a median.

In terms of motion-related factors, Alambeigi et al. (2020) found that the AV crash narratives include five topics: (1) transitions of control; (2) collisions at intersections in a right-turn lane; (3) collisions at intersections in non-turn lanes; (4) side-swipe crashes during a left-overtake; and (5) collisions at intersections involving oncoming traffic (Alambeigi et al., 2020). Kutela et al. (2022b) also found that the movements of AVs (i.e., turning, slowing down, stopping) are the key predictors of VRUs-AV-related crashes. Besides, Liu et al. (2021) summarized three important causes for AV crashes: (1) the perception-reaction time difference between AVs and human drivers; (2) AV's inaccurate identification of the intention of other vehicles to change lanes; and (3) AV's insufficient path planning in both spatial and temporal dimensions.

Interestingly, Ashraf et al. (2021) found that all types of AV crashes occurred mainly due to HVs. For example, rear-end crashes occurred because HVs failed to react appropriately to leading stopped AVs at intersections, while broadside, side-swipe, and head-on crashes occurred because of HVs' unsafe movements, red-light running, or disobeying the stop sign at the intersection.

To summarize, previous studies have identified various factors associated with AV crashes, and these factors could be categorized into two categories: (1) environment-related factors, which include roadway types (e.g., crosswalks, intersections, parking lots, two-way roads without a median), traffic signals, weather, and lighting; (2) motion-related factors, which include various movements of AVs (e.g., turning, slowing down, and stopping).

5.2.2.4 Factors Influencing AV Crash Outcomes

The injury severity level of road agents is a commonly adopted metric of crash outcomes (e.g., Abdel-Aty and Keller, 2005; Kockelman and Kweon, 2002; Savolainen et al., 2011). Thus, previous research also analyzed factors associated with the injury severity level in AV-involved crashes, from which we categorized the factors associated with AV crash severity into three categories: collision factor, AV factor, and environmental factor.

In terms of collision factor, rear-end collision was found to be associated with higher crash severity. For example, by using the property-damage-only (PDO) crash as the reference level, Xu et al. (2019) found that the severity level of rear-end crashes was higher than that of other types of collisions. Similarly, based on the Bayesian latent class model, Das et al. (2020) identified six classes of AV crash patterns, in which the class associated with rear-end collisions was also associated with a higher proportion of injury severity level (i.e., the percentage of the operator severity is more than 5 times higher than no severity condition). In addition, Das et al. (2020) also found that multi-vehicle collisions were associated with a higher proportion of injury severity level (i.e., the percentage of the operator severity is more than 2 times higher than no severity condition).

For the AV factor, both Xu et al. (2019) and Boggs et al. (2020) found that an increase in crash severity was positively associated with roadside parking. Das et al. (2020) also found that the AV crash pattern class associated with turning was associated with a higher proportion of injury severity level. Besides, irrespective of the collision type, when the AV was at fault, the chance of physical injury in a crash increased significantly (Kutela et al., 2022a; Wang and Li, 2019).

In terms of environment-related factors, Xu et al. (2019) found that AV crashes that occurred on one-way roads were more likely to be PDO crashes compared with AV crashes that occurred on other more complex roads. Chen et al. (2020) revealed that crash severity significantly increased when the AVs collided at an intersection under extreme weather conditions (e.g., fog and snow). Compared to areas where land-use patterns were less diverse, an AV crash resulting in injuries also had a higher likelihood of occurring in areas where land-use patterns were more diverse. Kutela et al. (2022a) also found that the likelihood of physical injury is higher at high speed, intersections, and wet pavements. Similarly, Yan et al. (2023a) found that a wet road surface significantly increased the probability of injury in AV crashes. Wang and Li (2019) also identified the highway as the location where severe injuries were likely to happen. Based on the Bayesian latent class model, Das et al. (2020) identified six classes of AV crash patterns, in which classes associated with dark lighting conditions with streetlights were also associated with a higher proportion of injury severity level compared to other AV crash pattern classes. Boggs et al. (2020) found that injury-involved crashes were positively associated with roadside parking, intersections, and arterial roads, while Lee et al. (2023) found that injury-involved crashes were more likely to occur on a ramp or slip lane. Interestingly, Liu et al. (2024) found that the mechanisms for AV crashes with different severities were different. For example, no damage and minor injured crashes were more likely to be associated with the road features (e.g., road surface conditions) and traffic conditions (e.g., intersection controls), while severe damage crashes are more likely to occur in a chain of events involving poor visibility, slippery road surface, and fixed objects.

TABLE 5.3
Factors associated with high injury severity levels in AV-involved crashes

Factor category	Factors associated with high injury severity levels	Source
Collision factor	Multi-vehicle collisions	Das et al. (2020)
	Rear-end collisions	Boggs et al. (2020); Das et al. (2020); Xu et al. (2019)
AV factor	Turning	Das et al. (2020)
	Roadside parking	Boggs et al. (2020); Xu et al. (2019)
	AV at fault	Kutela et al., (2022a); Wang and Li, (2019)
Environment-related factor	Wet pavements	Kutela et al., (2022a); Liu et al., (2024); Yan et al. (2023a)
	High-speed locations	Kutela et al., (2022a); Wang and Li, (2019)
	Intersections	Boggs et al. (2020); Chen et al. (2020)
	Extreme weather conditions (e.g., fog and snow)	Chen et al. (2020)
	Complex roads, diverse lane-use patterns	Chen et al. (2020); Xu et al. (2019)
	Dark lighting	Das et al. (2020)
	Arterial roads	Boggs et al. (2020)
	Ramp or slip lane	Lee et al. (2023)
	Fixed objects	Liu et al. (2024)

Table 5.3 summarizes the factors that were associated with high injury severity levels in AV-involved crashes. In all, existing studies have revealed several types of risky factors that were associated with high crash severity, which include collision factor (e.g., rear-end collision), AV factor (e.g., AV at fault), and environmental factor (e.g., road, weather, lighting). These factors could be considered valuable references when designing relevant countermeasures for enhancing mixed traffic safety.

5.3 PROACTIVE SAFETY ANALYSIS OF MIXED TRAFFIC

5.3.1 Key Scenarios

Three key scenarios were investigated in the mixed traffic in existing studies: (1) car-following scenario; (2) lane-changing scenario; and (3) car-passing scenario. Figure 5.1 presents these three scenarios.

As a vehicle's longitudinal movement in the surface transportation system, the car-following scenario was the most common and the most representative in the traffic safety analysis, typically including a leading vehicle and a following vehicle on the road. In the mixed traffic domain, existing studies tried to analyze the car-following characteristics of HV-following-AV, HV-following-HV, and AV-following-HV, depending on different research purposes (e.g., Huang et al., 2022; Wen et al., 2022).

As one common scenario that involved the lateral movement of the vehicle, the lane-changing scenario usually involves complex interactions between the ego vehicle and other vehicles or road users on the adjacent lane, which was critical to mixed traffic safety. Existing studies categorized the lane-changing scenario into two types, that is, mandatory lane changing and discretionary lane changing (Yang and Koutsopoulos, 1996). Mandatory lane changing is a required task and must be conducted to reach a specific destination while discretionary lane changing is voluntary and usually carried out to improve the current driving conditions. Some studies have explored discretionary lane changing given its importance in mixed traffic safety (e.g., Ali et al., 2024; Wen et al., 2023).

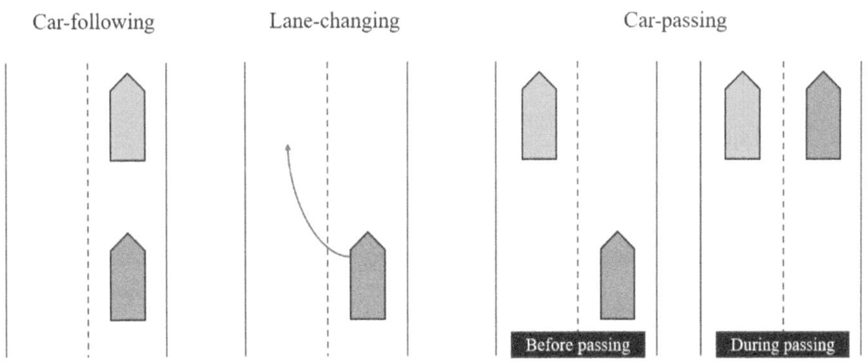

FIGURE 5.1 Three key scenarios in mixed traffic: Car-following, Lane-changing, and Car-passing.

The car-passing scenario mainly describes the process where a vehicle is passing another vehicle from the adjacent lane, which typically happens when a vehicle is trying to overtake another vehicle (e.g., Huang et al., 2022). For the approaches used for mixed traffic behavioral analysis, refer to Chapter 2 for more information.

5.3.2 Surrogate Safety Measures and Metrics

Although the crash analysis contributed a lot of valuable information to mixed traffic safety, the crashes were still rare events and thus distributed in a very sparse manner during the testing and deployment of AVs in mixed traffic. For mixed traffic safety, it would take a long time for AV safety impacts to be revealed in real-world crash data. Therefore, evaluating mixed traffic safety based on historical crash data is always post-hoc. To address this issue, surrogate safety measures (SSM) derived from traffic conflicts was proposed, as traffic conflicts were more frequent and can be predictive of crashes. Typically, traffic conflicts in mixed traffic were observable non-crash events, in which the interactions of HVs, AVs, and other road users construct a risk of collision if these road users continue to move in the original trajectories (Wang et al., 2021). According to Tarko (2018), there are two qualifying criteria for SSM: (1) it should be derived from traffic conflicts that are directly linked to crashes, and (2) the relationship between traffic conflicts and the related potential crash frequency and/or severity can be quantified by some practical methods. SSM acted as the important basis for the proactive safety analysis of mixed traffic.

The SSM metrics used by previous studies could be categorized into three categories, which include time-based SSM (e.g., TTC, time-exposed TTC, time-integrated TTC, time headway), deceleration-based SSM (e.g., deceleration rate to avoid the crash (DRAC), crash potential index (CPI), rear-end collision risk index (RCRI)), and energy-based SSM (e.g., DeltaV, conflict severity, crash index, conflict index) (Wang et al., 2021). Interested readers could refer to Wang et al. (2021) for a more comprehensive review of these SSM metrics.

However, most existing studies using SSM to evaluate mixed traffic safety were based on simulation (e.g., El-Hansali et al., 2021; Morando et al., 2018; Tafidis and Pirdavani, 2023), which were out of the scope of this chapter; thus we only included several studies that fitted our scope in the following sections.

5.3.3 Major Findings

5.3.3.1 Findings from Experimental Study

In terms of experimental studies, some existing studies found that AVs can enhance mixed traffic safety. For example, Wei and Shao (2024) investigated the effects of AVs on human drivers' following behaviors through a driving simulator study, in which they found that, compared to following conventional HDVs, drivers exhibited less driving volatility in speed and acceleration and longer TTC when following AVs. Through a field experiment, Mahdinia et al. (2021) found that AVs in mixed traffic streams can induce behavioral changes in conventional vehicle drivers, with some beneficial effects on safety (i.e., lower driving volatility and higher TTC when following an AV) and the environment (i.e., consume less fuel and produce fewer emissions when following an AV).

Another experimental study highlighted the potential influence of drivers' subjective differences. Zhao et al. (2020) also investigated the differences between HV-following-HV and HV-following-AV using a field experiment, and they found that different drivers exhibit different behaviors in following-AV versus following-HV: some are AV-believers, some are AV-skeptics, and the others are insensitive. However, there was no significant difference between following a lead AV and following a lead HV, which indicated that drivers' response to the lead vehicle depends on their subjective trusts on AV technologies rather than the actual driving behavior.

5.3.3.2 Findings from Analysis of Real-World Data

In terms of analysis of real-world data, in line with the findings from experimental studies, most studies found that human drivers interacted with AVs more safely compared to when interacted with HVs. Many studies were conducted based on the Waymo Open Dataset. For example, Huang et al. (2022) explored human drivers' interactions with AVs in two types of events on highways, that is, car-following events and car-passing events. They found that human drivers tended to keep a larger safety margin with increasing speed regardless of the types of vehicles they interacted with, by keeping a larger gap distance, and a larger TTC to the lead vehicle in car-following events. At the same time, drivers kept a larger lateral distance to AVs in car-passing events regardless of the lead vehicle's speed. Similarly, by using the Waymo Open Dataset, Wen et al. (2022) characterized human drivers' following behaviors and compared HV-following-HV with HV-following-AV in terms of several safety metrics (e.g., TTC), they found that human drivers had lower driving volatility in terms of velocity and acceleration/deceleration, smaller time headways and higher TTC. Hu et al. (2023) also tried to characterize the empirical car-following behaviors of AVs and compare AVs' car following behaviors with HVs' car following behaviors. They found that AVs were much safer than HVs based on several SSM, as time headways and jam spacings of the AVs were significantly larger than HVs. The authors also noted that the current AV behaved in a conservative way to ensure its safety at the cost of efficiency. Ali et al. (2024) investigated the discretionary lane-changing execution behavior of AVs and compared AVs' behavior with HVs, and they found that AVs maintain larger lead and lag gaps, longer discretionary lane-changing execution time, and lower acceleration variation than HVs, which means that AV presented safer discretionary lane-changing behavior compared to HVs (Ali et al., 2024). Based on the Lyft Level-5 dataset, Li et al. (2023) also compared the operation behaviors of HV when following different leading vehicles, that is, HV-following-AV and HV-following-HV. They found that mixed traffic safety had been enhanced after the introduction of AV, which was evidenced by the changes in HV behaviors (i.e., less aggressive, larger TTC).

To summarize, existing studies using real-world data revealed that human drivers seem to interact with AVs in a generally safer way, compared to interacting with HVs. However, more research is still needed as the data used in existing studies were still limited and might be biased in representing human drivers' behaviors in mixed traffic.

5.4 POTENTIAL RESEARCH DIRECTIONS FOR IMPROVING MIXED TRAFFIC SAFETY

As suggested by findings from crash analysis and behavioral analysis in existing empirical studies, we proposed several potential research directions for improving mixed traffic safety as follows:

- AVs need to be improved in terms of detecting and warning human drivers of the following vehicles approaching from behind to avoid rear-end collisions. For example, AVs could communicate the current status and intention with other HVs through external human–machine interfaces (eHMI) (e.g., Dey et al., 2021; Papakostopoulos et al., 2021; Yan et al., 2023b). Until now, there were no established standards on how to design the eHMI for AVs and what information should the eHMI include, future research may explore the effectiveness of different eHMI designs (e.g., appearance, included information).
- To reduce injury crashes, the surface transportation systems should improve the roadway infrastructures to better support AVs. For example, one possible solution for avoiding human drivers' rear-ended AVs could be to separate AVs from HVs by using AV-only lanes (Yu et al., 2019). However, there might be a trade-off between safety and efficiency, future research could explore the effects of different roadway infrastructures on the safety and efficiency of the whole transportation system.
- The risky factors of AV crashes identified from existing studies could be used in several ways: (1) to develop safety-critical testing scenarios for efficient and reasonable testing for future driving automation (Kibalama et al., 2022; Lee et al., 2024; Song et al., 2021); (2) to help develop V2V and vehicle-to-infrastructure (V2I) technologies (Harding et al., 2014); and (3) to guide the design and optimization of AV algorithms in specific scenarios (Sharma et al., 2021).
- Inspired by drone-based and computer vision-based techniques (Zheng et al., 2024), we highly recommend research efforts on collecting the trajectory data that includes the interaction between AVs and HVs on publicly open roads and analyzing the mixed traffic safety through SSM.

REFERENCES

Abdel-Aty, M., Keller, J., 2005. Exploring the overall and specific crash severity levels at signalized intersections. *Accid. Anal. Prev.* 37 3, 417–425.

Alambeigi, H., McDonald, A.D., Tankasala, S.R., 2020. Crash themes in automated vehicles: A topic modeling analysis of the California Department of Motor Vehicles automated vehicle crash database. *arXiv Prepr.* arXiv2001.11087.

Ali, Y., Sharma, A., Chen, D., 2024. Investigating autonomous vehicle discretionary lane-changing execution behaviour: Similarities, differences, and insights from Waymo dataset. *Anal. Methods Accid. Res.* 42, 100332. doi:10.1016/j.amar.2024.100332

Arvin, R., Khattak, A.J., Kamrani, M., Rio-Torres, J., 2021. Safety evaluation of connected and automated vehicles in mixed traffic with conventional vehicles at intersections. *J. Intell. Transp. Syst.* 25 2, 170–187. doi:10.1080/15472450.2020.1834392

Ashraf, M.T., Dey, K., Mishra, S., Rahman, M.T., 2021. Extracting rules from autonomous-vehicle-involved crashes by applying decision tree and association rule methods. *Transp. Res. Rec. J. Transp. Res. Board* 2675 11, 522–533. doi:10.1177/03611981211018461

Banerjee, S.S., Jha, S., Cyriac, J., Kalbarczyk, Z.T., Iyer, R.K., 2018. Hands off the wheel in autonomous vehicles?: A systems perspective on over a million miles of field data, in: *2018 48th Annual IEEE/IFIP International Conference on Dependable Systems and Networks (DSN)*. IEEE, pp. 586–597. doi:10.1109/DSN.2018.00066

Blanco, M., Atwood, J., Russell, S.M., Trimble, T.E., McClafferty, J.A., Perez, M.A., 2016. *Automated Vehicle Crash Rate Comparison Using Naturalistic Data*. Virginia Tech Transportation Institute.

Boggs, A.M., Wali, B., Khattak, A.J., 2020. Exploratory analysis of automated vehicle crashes in California: A text analytics & hierarchical Bayesian heterogeneity-based approach. *Accid. Anal. Prev.* 135, 105354. doi:10.1016/j.aap.2019.105354

Chen, H., Chen, H., Liu, Z., Sun, X., Zhou, R., 2020. Analysis of factors affecting the severity of automated vehicle crashes using XGBoost model combining POI data. *J. Adv. Transp.* 2020, 1–12. doi:10.1155/2020/8881545

Das, S., Dutta, A., Tsapakis, I., 2020. Automated vehicle collisions in California: Applying Bayesian latent class model. *IATSS Res.* 44 4, 300–308. doi:10.1016/j.iatssr.2020.03.001

Dey, D., Matviienko, A., Berger, M., Pfleging, B., Martens, M., Terken, J., 2021. Communicating the intention of an automated vehicle to pedestrians: The contributions of eHMI and vehicle behavior. *IT – Informat. Technol.* 63 2, 123–141.

Dixit, V.V., Chand, S., Nair, D.J., 2016. Autonomous vehicles: Disengagements, accidents and reaction times. *PLoS One* 11 12, e0168054. doi:10.1371/journal.pone.0168054

El-Hansali, Y., Farrag, S., Yasar, A., Shakshuki, E., Al-Abri, K., 2021. Using surrogate measures to evaluate the safety of autonomous vehicles. *Procedia Comput. Sci.* 191, 151–159. doi:10.1016/j.procs.2021.07.020

Favarò, F.M., Nader, N., Eurich, S.O., Tripp, M., Varadaraju, N., 2017. Examining accident reports involving autonomous vehicles in California. *PLoS One* 12 9, e0184952. doi:10.1371/journal.pone.0184952

Garg, M., Bouroche, M., 2023. Can connected autonomous vehicles improve mixed traffic safety without compromising efficiency in realistic scenarios? *IEEE Trans. Intell. Transp. Syst.* 24 6, 6674–6689. doi:10.1109/TITS.2023.3238889

Goodall, N., 2024. Normalizing crash risk of partially automated vehicles under sparse data. *J. Transp. Saf. Secur.* 16 1, 1–17. doi:10.1080/19439962.2023.2178566

Goodall, N.J., 2021. Comparison of automated vehicle struck-from-behind crash rates with national rates using naturalistic data. *Accid. Anal. Prev.* 154, 106056. doi:10.1016/j.aap.2021.106056

Harding, J., Powell, G., Yoon, R., Fikentscher, J., Doyle, C., Sade, D., Lukuc, M., Simons, J., Wang, J., 2014. *Vehicle-to-Vehicle Communications: Readiness of V2V Technology for Application*. United States: National Highway Traffic Safety Administration.

Hu, X., Zheng, Z., Chen, D., Sun, J., 2023. Autonomous vehicle's impact on traffic: Empirical evidence from Waymo open dataset and implications from modelling. *IEEE Trans. Intell. Transp. Syst.* 24 6, 6711–6724. doi:10.1109/TITS.2023.3258145

Huang, C., He, D., Wen, X., Yan, S., 2023a. Beyond adaptive cruise control and lane centering control: Drivers' mental model of and trust in emerging ADAS technologies. *Front. Psychol.* 14. doi:10.3389/fpsyg.2023.1236062

Huang, C., Wang, A., Yan, S., He, D., 2024. Investigating the interrelationships among factors associated with automated vehicle crashes using additive Bayesian network. *Transp. Res. Rec. J. Transp. Res. Board*. doi:10.1177/03611981241274152

Huang, C., Wen, X., He, D., 2023b. Characteristics of rear-end collisions: A comparison between automated driving system-involved crashes and advanced driving assistance system-involved crashes. *Transp. Res. Rec. J. Transp. Res. Board*, 2678 7. doi:10.1177/03611981231209319

Huang, C., Wen, X., He, D., Jian, S., 2022. Sharing the road: How human drivers interact with autonomous vehicles on highways. *Proc. Hum. Factors Ergon. Soc. Annu. Meet.* 66 1, 1437–1441. doi:10.1177/1071181322661165

Jiang, Y., Ren, T., Ma, Y., Wu, Y., Yao, Z., 2023. Traffic safety evaluation of mixed traffic flow considering the maximum platoon size of connected automated vehicles. *Phys. A Stat. Mech. Appl.* 612, 128452. doi:10.1016/j.physa.2023.128452

Kibalama, D., Tulpule, P., Chen, B.-S., 2022. *AV/ADAS Safety-Critical Testing Scenario Generation from Vehicle Crash Data.* SAE Technical Paper.

Kockelman, K.M., Kweon, Y.-J., 2002. Driver injury severity: An application of ordered probit models. *Accid. Anal. Prev.* 34 3, 313–321.

Kutela, B., Avelar, R.E., Bansal, P., 2022a. Modeling automated vehicle crashes with a focus on vehicle at-fault, collision type, and injury outcome. *J. Transp. Eng. Part A Syst.* 148 6. doi:10.1061/JTEPBS.0000680

Kutela, B., Das, S., Dadashova, B., 2022b. Mining patterns of autonomous vehicle crashes involving vulnerable road users to understand the associated factors. *Accid. Anal. Prev.* 165, 106473. doi:10.1016/j.aap.2021.106473

Lee, H., Kang, M., Hwang, K., Yoon, Y., 2024. The typical AV accident scenarios in the urban area obtained by clustering and association rule mining of real-world accident reports. *Heliyon* 10 3, e25000. doi:10.1016/j.heliyon.2024.e25000

Lee, S., Arvin, R., Khattak, A.J., 2023. Advancing investigation of automated vehicle crashes using text analytics of crash narratives and Bayesian analysis. *Accid. Anal. Prev.* 181, 106932. doi:10.1016/j.aap.2022.106932

Li, T., Han, X., Ma, J., Ramos, M., Lee, C., 2023. Operational safety of automated and human driving in mixed traffic environments: A perspective of car-following behavior. *Proc. Inst. Mech. Eng. Part O J. Risk Reliab.* 237 2, 355–366. doi:10.1177/1748006X211050696

Liu, P., Guo, Y., Liu, P., Ding, H., Cao, J., Zhou, J., Feng, Z., 2024. What can we learn from the AV crashes? – An association rule analysis for identifying the contributing risky factors. *Accid. Anal. Prev.* 199, 107492. doi:10.1016/j.aap.2024.107492

Liu, Q., Wang, X., Wu, X., Glaser, Y., He, L., 2021. Crash comparison of autonomous and conventional vehicles using pre-crash scenario typology. *Accid. Anal. PRev.* 159, 106281. doi:10.1016/j.aap.2021.106281

Mahdinia, I., Mohammadnazar, A., Arvin, R., Khattak, A.J., 2021. Integration of automated vehicles in mixed traffic: Evaluating changes in performance of following human-driven vehicles. *Accid. Anal. Prev.* 152, 106006. doi:10.1016/j.aap.2021.106006

McCarthy, R.L., 2022. Autonomous vehicle accident data analysis: California OL 316 reports: 2015–2020. *ASCE-ASME J Risk Uncert Engrg Sys Part B Mech Engrg* 8 3. doi:10.1115/1.4051779

Morando, M.M., Tian, Q., Truong, L.T., Vu, H.L., 2018. Studying the safety impact of autonomous vehicles using simulation-based surrogate safety measures. *J. Adv. Transp.* 2018, 1–11. doi:10.1155/2018/6135183

Novat, N., Kidando, E., Kutela, B., Kitali, A.E., 2023. A comparative study of collision types between automated and conventional vehicles using Bayesian probabilistic inferences. *J. Safety Res.* 84, 251–260. doi:10.1016/j.jsr.2022.11.001

Papakostopoulos, V., Nathanael, D., Portouli, E., Amditis, A., 2021. Effect of external HMI for automated vehicles (AVs) on drivers' ability to infer the AV motion intention: A field experiment. *Transp. Res. Part F Traffic Psychol. Behav.* 82, 32–42.

Petrović, Đ., Mijailović, R., Pešić, D., 2020. Traffic accidents with autonomous vehicles: Type of collisions, manoeuvres and errors of conventional vehicles' drivers. *Transp. Res. Procedia* 45, 161–168. doi:10.1016/j.trpro.2020.03.003

Regev, S., Rolison, J.J., Moutari, S., 2018. Crash risk by driver age, gender, and time of day using a new exposure methodology. *J. Safety Res.* 66, 131–140. doi:10.1016/j.jsr.2018.07.002

Savolainen, P.T., Mannering, F.L., Lord, D., Quddus, M.A., 2011. The statistical analysis of highway crash-injury severities: A review and assessment of methodological alternatives. *Accid. Anal. Prev.* 43 5, 1666–1676.

Schoettle, B., Sivak, M., 2015. *A Preliminary Analysis of Real-World Crashes Involving Self-Driving Vehicles*. University of Michigan Transportation Research Institute.

Schwall, M., Daniel, T., Victor, T., Favaro, F., Hohnhold, H., 2020. Waymo public road safety performance data. *arXiv Prepr.* arXiv2011.00038.

Sharma, O., Sahoo, N.C., Puhan, N.B., 2021. Recent advances in motion and behavior planning techniques for software architecture of autonomous vehicles: A state-of-the-art survey. *Eng. Appl. Artif. Intell.* 101, 104211.

Song, Y., Chitturi, M.V., Noyce, D.A., 2021. Automated vehicle crash sequences: Patterns and potential uses in safety testing. *Accid. Anal. Prev.* 153, 106017. doi:10.1016/j.aap.2021.106017

Tafidis, P., Pirdavani, A., 2023. Application of surrogate safety measures in higher levels of automated vehicles simulation studies: A review of the state of the practice. *Traffic Inj. Prev.* 24 3, 279–286. doi:10.1080/15389588.2023.2176711

Tarko, A.P., 2018. Surrogate measures of safety, in: *Safe Mobility: Challenges, Methodology and Solutions*. Emerald Publishing Limited, pp. 383–405.

Teoh, E.R., Kidd, D.G., 2017. Rage against the machine? Google's self-driving cars versus human drivers. *J. Safety Res.* 63, 57–60. doi:10.1016/j.jsr.2017.08.008

Wang, C., Xie, Y., Huang, H., Liu, P., 2021. A review of surrogate safety measures and their applications in connected and automated vehicles safety modeling. *Accid. Anal. Prev.* 157, 106157. doi:10.1016/j.aap.2021.106157

Wang, S., Li, Z., 2019. Exploring the mechanism of crashes with automated vehicles using statistical modeling approaches. *PLoS One* 14 3, e0214550. doi:10.1371/journal.pone.0214550

Wei, S., Shao, M., 2024. Existence of connected and autonomous vehicles in mixed traffic: Impacts on safety and environment. *Traffic Inj. Prev.* 25 3, 390–399. doi:10.1080/15389588.2023.2291337

Wen, X., Cui, Z., Jian, S., 2022. Characterizing car-following behaviors of human drivers when following automated vehicles using the real-world dataset. *Accid. Anal. Prev.* 172, 106689. doi:10.1016/j.aap.2022.106689

Wen, X., Huang, C., Jian, S., He, D., 2023. Analysis of discretionary lane-changing behaviours of autonomous vehicles based on real-world data. *Transp. A Transp. Sci.* 1–24.

Xu, C., Ding, Z., Wang, C., Li, Z., 2019. Statistical analysis of the patterns and characteristics of connected and autonomous vehicle involved crashes. *J. Safety Res.* 71, 41–47. doi:10.1016/j.jsr.2019.09.001

Yan, S., Huang, C., He, D., 2023a. A comparison of patterns and contributing factors of ADAS and ADS involved crashes. *J. Transp. Saf. Secur.* 1–28. doi:10.1080/19439962.2023.2284175

Yan, S., Huang, C., Xie, W., He, D., 2023b. HMI design for chain-braking event based on V2V communication, in: *Adjunct Proceedings of the 15th International Conference on Automotive User Interfaces and Interactive Vehicular Applications*. pp. 105–110.

Yang, Q., Koutsopoulos, H.N., 1996. A microscopic traffic simulator for evaluation of dynamic traffic management systems. *Transp. Res. Part C Emerg. Technol.* 4 3, 113–129. doi:10.1016/S0968-090X(96)00006-X

Yao, Z., Hu, R., Jiang, Y., Xu, T., 2020. Stability and safety evaluation of mixed traffic flow with connected automated vehicles on expressways. *J. Safety Res.* 75, 262–274. doi:10.1016/j.jsr.2020.09.012

Ye, L., Yamamoto, T., 2019. Evaluating the impact of connected and autonomous vehicles on traffic safety. *Phys. A Stat. Mech. Appl.* 526, 121009. doi:10.1016/j.physa.2019.04.245

Yu, H., Tak, S., Park, M., Yeo, H., 2019. Impact of autonomous-vehicle-only lanes in mixed traffic conditions. *Transp. Res. Rec. J. Transp. Res. Board* 2673 9, 430–439. doi:10.1177/0361198119847475

Zhao, X., Wang, Z., Xu, Z., Wang, Y., Li, X., Qu, X., 2020. Field experiments on longitudinal characteristics of human driver behavior following an autonomous vehicle. *Transp. Res. Part C Emerg. Technol.* 114, 205–224. doi:10.1016/j.trc.2020.02.018

Zheng, O., Abdel-Aty, M., Yue, L., Abdelraouf, A., Wang, Z., Mahmoud, N., 2024. CitySim: A drone-based vehicle trajectory dataset for safety-oriented research and digital twins. *Transp. Res. Rec. J. Transp. Res. Board* 2678 4, 606–621. doi:10.1177/03611981231185768

6 Environmental Impact of the Mixed Traffic and Countermeasures

Ran Tu and Shan Xue

6.1 ENERGY AND EMISSIONS: THE MEASURES OF ENVIRONMENTAL IMPACT OF ROAD TRAFFIC

This section gives a brief introduction to traffic energy consumption and emissions regarding the source and the estimation methods.

6.1.1 Introduction to the Source of Energy Consumption and Emissions

Energy consumption and emissions are probably the most recognized and studied metrics for the environmental impact of road traffic. The energy consumption of vehicles comes from the power demand for vehicle engines and accessories to move forward and work properly. For fossil fuel-powered vehicles with internal combustion engines, fuel combustion generates air pollutant and greenhouse gas (GHG) emissions (or tailpipe emissions); while for electric vehicles (EVs), emissions are shifted to the upstream electricity grid, and the source of electricity matters. Another important source of emissions is non-tailpipe emissions, which can be categorized into brake wear, tire wear, rad wear, and resuspension. Although vehicle electrification eliminates tailpipe emissions, non-tailpipe emissions do not necessarily decrease due to a higher vehicle weight. In addition, upstream emissions from the electricity grid largely depend on the energy source of the grid. Thus, electrification is not capable of mitigating traffic-related air pollution (TRAP) on its own. The debate on EVs will be discussed later in this chapter. This section will focus on energy consumption and tailpipe emissions.

6.1.2 Estimation Methods, Their Applicable Scopes, and Data Requirements

Traffic energy consumption and emissions can be described in three scopes: microscopic from the vehicle perspective, mesoscopic from the road perspective, and macroscopic from the network or regional perspective. Corresponding to the scope of the modeling, the estimation methods can be categorized into two types: top-down and bottom-up. The top-down method, generally only applicable for macroscopic

DOI: 10.1201/9781032674667-6

studies, is widely applied to estimate regional or national traffic energy consumption and emissions (Coelho et al., 2014). Under the framework of SNAP (Selected Nomenclature Air Pollution, such as combustion in energy and transformation industries, production processes, and agriculture), the top-down method can be used for traffic emissions in compliance with SNAP 7.

As the name indicates, the top-down method first calculates total energy consumption from a regional/national scope using vehicle registration data, average vehicle-kilometer-traveled from surveys, and the energy consumption factor for each unit of traffic activity. Similar to energy consumption, the top-down method calculates emissions from either an aggregated emission factor or a conversion factor from energy consumption to a certain type of pollutant. The regional result is then disaggregated from the total value to the grid level based on the spatial characteristics, such as street density and population percentage (de Eicker et al., 2008; Fameli & Assimakopoulos, 2015; Progiou & Ziomas, 2011). The top-down method is advantageous in regional planning research owing to its simple computation and data requirement.

The top-down method cannot differentiate link-specific traffic conditions, which strongly affect energy consumption and emissions from meso- and microscopes. The bottom-up approach, which applies link-based or vehicle-specific traffic data as input, is preferred in this circumstance. The bottom-up approach should be first applied to road-specific and individual vehicles, then aggregate to the network wide if needed.

The input data for the bottom-up method should include more details compared to that of the top-down method. Traffic activity data for the bottom-up approach can be generated from traffic simulation or real-time traffic sensors such as roadside cameras, radar, and vehicle positioning systems. For the road-specific level, average traffic conditions, such as average speed, volume, and density, are the commonly adopted variables to describe the traffic activity (Colberg et al., 2005; Luo et al., 2018; Tanvir, 2018). In addition, the energy or emission factors (in grams or joule per unit distance) can be generated from local databases, such as Calculations of Emissions from Road Transport (COPERT) and Handbook Emission Factors for Road Transport (HBEFA) (Chen & Borken-Kleefeld, 2014; Kawsar et al., 2024), the two most widely used databases in Europe. For the individual vehicle level, instantaneous operations, such as second-by-second speed and acceleration, are taken as inputs, and the associated energy consumption or emission rate (in grams/joule per unit time) is either estimated by a regression model or retrieved from a database. Unlike models for road-specific levels, individual vehicle-based models generate more accurate estimations, but they require a longer computational process and more precise traffic data. Therefore, a trade-off should be considered when selecting appropriate estimation models for specific research objectives.

Multiple energy consumption and emission models have been developed worldwide. The models can be categorized into physical models and regression-based models according to their modeling methods and frameworks. Physical models refer to the methods that are built upon the power demand of vehicles to move forward. The fuel/electricity consumption and emissions can then be estimated based on the conversion factor or the experimental database. For example, MOtor (USEPA) is widely applied in North America. In MOVES, base emission rates are assigned based on vehicle operating mode (OpMode), which can be determined by combination bins of speed and vehicle-specific power (VSP, equation 6.1), and adjusted with vehicle types and meteorological conditions (U.S. Environmental Protection Agency, 2023):

$$P_{V,t} = \frac{Av_t + Bv_t^2 + Cv_t^3 + mv_t a_t}{m} \tag{6.1}$$

Where
$P_{V,t}$ is the vehicle-specific power, or VSP;
v_t is the speed at time t (m/s);
a_t is the acceleration at time t (m/s^2);
m is the mass of the vehicle, usually referred to as "weight" (Mg)。

A, B, and C are track-road coefficients, representing rolling resistance, rotational resistance, and aerodynamic drag in unit $kW - sec/m$, $kW - sec^2/m^2$, and $kW - sec^3/m^3$.

Another example of physical modeling is the Comprehensive Modal Emission Model (CMEM, Center for Environmental Research and Technology, 2017), which was developed by the Center for Environmental Research and Technology at the University of California, Riverside. The model calculates fuel consumption rate and emissions from internal vehicle operation data, including engine efficiency and engine power. The use of internal parameters elevates the accuracy of the model (Frey et al., 2010).

As compared to physical modeling, regression-based energy/emission models are widely developed with external vehicle operations (including speed and acceleration) as the input to estimate vehicle energy usage and emissions. Among them, VT-MICRO and Virginia Tech Comprehensive Power-based Fuel Consumption Model (VT-CPFM), developed by the Virginia Tech Transportation Institute, are recognized examples (Almannaa et al., 2019; Rakha et al., 2004). Due to the regenerative braking process of EVs, the energy consumption model for EVs needs an additional step to distinguish between energy regeneration and depletion. The differentiation can be determined by the value of acceleration by simply assuming that a negative acceleration value refers to the regenerative braking or through a well-trained model, which uses the drive cycle as the input and the regenerative state as the output. Figure 6.1 illustrates the training process of the EV energy consumption model.

As the figure shows, the power output of the EV can be fit using a regression model, which is presented through equation 6.2 to equation 6.5. The regeneration state, where the value of the power is negative, is identified only when the deceleration occurs.

$$P_{Wheel}(t) = (ma(t) + mg\cos(\theta)\frac{C_r}{1000}(c_1 v(t) + c_2) + \frac{1}{2}\rho_{air} A_f C_D v^2(t) + mg\sin(\theta))v(t) \tag{6.2}$$

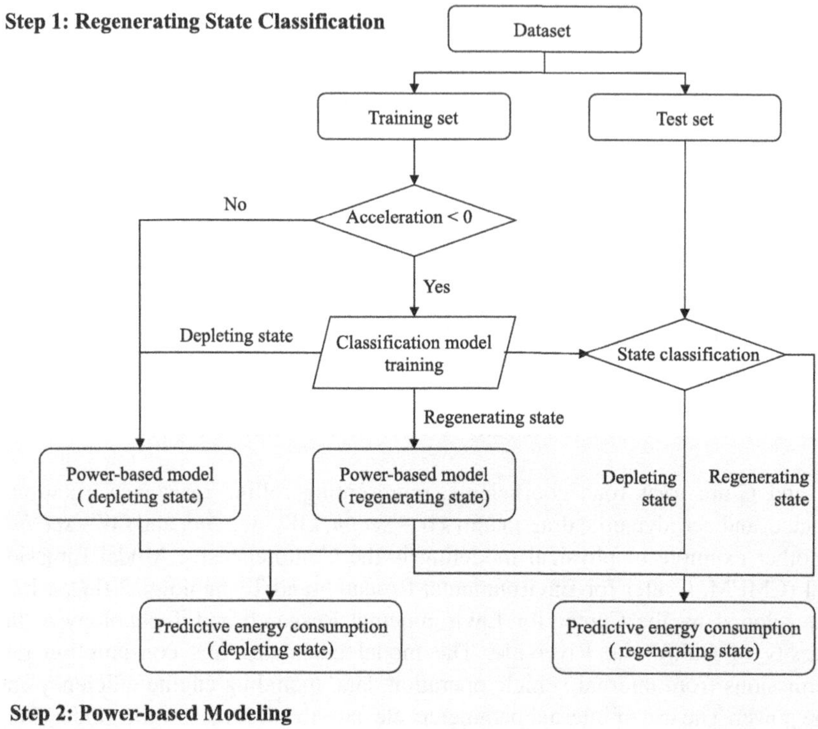

FIGURE 6.1 EV energy consumption modeling framework with a regeneration identifier (Xu et al., 2023).

Where

P_{Wheel} is wheel power of the vehicle;

v_t is the speed at time t (m/s); a_t is the acceleration at time t (m/s^2); m is the mass of the vehicle, (m);

θ is the road grade;

C_r, c_1, c_2 are the rolling resistance parameters representing the road surface type, road condition, and vehicle tire type, respectively;

ρ_{Air}, A_f, and $C_D$represent air density (kg/m^3), the front area of the vehicle m_2, and the aerodynamic drag coefficient of the vehicle, respectively.

The power need of the vehicle is proportional to P_{wheel}, as equation 6.3, and η represents the combination of motor efficiency, driveline efficiency, and other associated energy loss.

$$P_{Wheel}(t) = P(t)\,\eta \tag{6.3}$$

Overall, the instantaneous energy consumption of the EV is equation 6.4, which can be further simplified as equation 6.5.

$$E(t) = \frac{P(t)}{3600 \times 1000} \tag{6.4}$$

$$\begin{aligned} E(t) = \beta_0 + \beta_1 v(t)\sin(\theta) + \beta_2 v(t)\cos(\theta) + \beta_3 v^2(t)\cos(\theta) \\ + \beta_4 v^3(t) + \beta_5 a(t) v(t) \end{aligned} \tag{6.5}$$

6.2 ENVIRONMENTAL IMPACTS OF MIXED TRAFFIC: HOW IT DISTINGUISHES FROM NORMAL TRAFFIC

The impact of CAVs can be reflected through the total vehicle kilometer traveled (VKT), which can be influenced by the induced demand from introducing CAVs, and different routing strategies of CAVs. In addition, emissions and energy consumption associated with CAVs per VKT are significantly influenced by the distinguished driving behavior. Other influencing factors, such as the fuel type, the vehicle-to-everything (V2X) communication, and the market penetration rate (MPR), also affect total energy usage and emissions from the mixed traffic (Kopelias et al., 2020).

The evaluation of the CAV impacts on mixed traffic in terms of energy and emissions can be compared before-and-after using vehicle drive cycles. The energy/emissions are mostly estimated using the abovementioned models in Section 1.1, given limited access to the real-world measurement and access to the vehicle electronic control unit (ECU). In this section, environmental impact assessments through empirical data analysis and simulation-based experiments are reviewed, respectively.

6.2.1 Assessing the Performance of CAVs Using Empirical Data

Small-scale driving tests, which are usually implemented in a controlled environment in a certain length of expressway without little interference from other road users, have been practiced for the last decades. Several studies have demonstrated the energy-saving effects based on the vehicle operating data from these tests (Robinette et al., 2019). However, experiments with controlled road conditions do not consider interactions between CAVs and other road users, thus being unable to reveal the actual energy-saving and emission-reduction effects of on-road CAVs.

Recently, large-scale tests on open public roads, such as trials from Waymo, Apollo, and Pony.ai, as listed in Chapter 3 of the book, provided a huge amount of data for researchers to validate the benefit of CAVs. For example, Qu et al. (2022) shows that the connectivity can elevate the energy-saving effect by over 10% compared to the AV technology alone. Zhang et al. (2022) found the platooning of CAVs can reduce total energy consumption by 67%. Not only CAVs but the following human-driving vehicles in the platoon also exhibit fuel-saving potentials, as proved by Mahdinia et al. (2021) using the pilot test in Texas, U.S.

Given the positive environmental influence of CAV applications, a recent study in Shanghai revealed that although CAVs present energy-saving benefits in normal driving conditions, the disengagement of early-stage CAVs, a process of the human drivers taking control of the vehicle to avoid accidents, lead to higher energy usage than normal conditions by 60% (Tu et al., 2024). Disengagement is a necessary setup for CAVs in the current stage to ensure driving safety, and the excess energy consumption from this process highlights the environmental benefits of improving the sensing and control systems of CAVs.

Empirical experiments shed light on the environmental benefits of individual vehicles through improved driving behavior. However, due to limited data access to other road users, the effect on traffic flow cannot be properly evaluated. Nevertheless, recent datasets, such as the Waymo open dataset, provide high-resolution instantaneous records for both tested CAVs and the surrounding traffic, such as the heading and following vehicles, and the cyclists and pedestrians within the sensing range. The interaction between each pair of road users can be captured. For example, studies have been conducted to explore the car-following behavior of AVs and the difference between AVs and human vehicles (HVs) (Hu et al., 2022, 2023; Li et al., 2023). In addition, a compiled public dataset that is processed based on the public raw data gives easier access for researchers to identify differences between AVs and HVs in terms of the energy consumption characteristics (Hu et al., 2021; Chen et al., 2023). The result can be further applied to justify simulation parameters and, in turn, facilitate the representativeness of simulation experiments.

6.2.2 Assessing the Performance of CAVs through Simulation Experiments

In simulation-based experiments, the parameters of CAV operations need to be pre-justified. Simulation parameters that are tuned for the CAV experiment can be categorized into three groups. The first category specifies the vehicle model, including the size, the weight, and the fuel type. The second category describes the vehicle motion features, including the headway time, the reaction time, the maximum acceleration, and the minimum acceleration. The third category sets the traffic condition of the simulation, including the road geometry, the traffic volume, the signal control, and the MPR of vehicles that are embedded with the CAV control algorithm. A more detailed introduction to the modeling and settings of the CAV simulation can be found in Chapter 4 and Chapter 5 of the book.

The advantage of the simulation, compared to the empirical data experiments, is the scalability of the test. The effects of CAVs on energy consumption and emissions can be revealed from a microscopic individual vehicle level to a macroscopic network-wide level. In general, the application of CAVs demonstrates a positive influence on environmental improvement. However, results vary across different simulation settings and control algorithms.

Three common types of simulation domains can be summarized from the existing literature, including intersection, road corridors (Yao et al., 2021), and road networks (Tu et al., 2017; Stogios, 2018). Studies have observed that the mixed traffic flow becomes more stable with fewer collisions and a shorter traffic delay (Qin et al.,

2020a,b). As a co-benefit, the energy consumption of the mixed traffic can be reduced by up to 60%.

In addition to changing the car-following and reaction time of CAVs that have been widely adopted in previous studies (Stogios, 2018; Djavadian & Farooq, 2018), the driving behavior of HVs should also be calibrated. The empirical data with the detected information of surrounding road users can be an asset to the simulation calibration, as mentioned in Section 2.1.

Energy and emissions are influenced by vehicle operational characteristics through various mechanisms, such as driver reaction time, safe spacing distance, gap acceptance thresholds, and lane-changing behaviors (Atkins, 2016). For this reason, the performance of AVs heavily depends on the driving parameter settings, and it is one of the major concerns and biggest sources of result variation of simulation-based studies.

From the corridor and network perspectives, due to the impact of CAVs on the behavior of the following vehicle, introducing CAVs to the traffic can effectively leverage the features of traffic flow, leading to a systematic change. Therefore, researchers should be aware of not only the simulation parameters but also the induced demand from the CAV introduction. The network-level impact of the mixed traffic can be further read in Chapter 8.

Simulation experiments assume an ideal environment and a full execution of the control algorithm. However, the decision-making and manual intervention of immature AV technologies in complex traffic conditions are ignored. Given a rising trend of studies on compliance, the uncertainty of the human drivers, especially for the scenario of a human driver in a CAV, has seldom been explored. This results in an overestimation of the improvement of AV energy efficiency (Qu et al., 2022).

6.3 MITIGATING THE ENVIRONMENTAL IMPACTS BROUGHT BY CAVS

6.3.1 Reducing Energy and Emissions from a Smoother Driving Style

The application of CAVs leads to a disruptive change in the dynamics of traffic flow, affecting the energy and emissions that are generated by road vehicles. Theoretically, the introduction of CAVs effectively improves the stability and efficiency of road traffic, thus generating larger environmental benefits (Pan et al., 2024). However, given the elevated vehicle control, CAVs in practice are not necessarily associated with positive environmental impacts.

Given the developing eco-driving method, which aims to reduce fuel consumption and emissions through smoother driving behavior, current studies show that the influence is greatly differentiated according to traffic conditions and control strategies. For example, Garcia-Castro et al. (2017) found that the CAV application with longer headways and smoother acceleration can generate higher emissions under heavy traffic demand. In another study that was conducted on a network with mixed road types, the researcher found reduced air pollutant emissions (e.g., HC and CO) after the MPR of the eco-driving-enabled CAVs is over 60% on highway control, and the overall emission reduction can only be observed when the MPR reaches 80% (Tu et al., 2017).

Technologies on vehicle-to-infrastructure (V2I) enable the coordination between vehicles and signal control at intersections. The integration can be effectively used to reduce traffic delays, and as a bonus, vehicle emissions and energy consumption can also be mitigated. For example, Jiang et al. (2022) developed a multi-phase integrated control framework to control the longitudinal/latitudinal movement of vehicles and optimize the signal phase of the approaching intersection. The framework is able to reduce air pollutant emissions by up to 55.2% with up to 22.14% reduced energy usage. Similar findings can also be found in Li et al. (2024) and Di Pace et al. (2022), where a fuel reduction over 15% can be achieved. Even without the V2I technology, researchers proposed a delayed self-reinforcement (DSR) algorithm for the vehicle spacing control to elevate the behavior of AVs without mutual connection, and the result can be close to that of CAVs (Lin et al., 2023).

Besides, the distributed control method, which is facilitated by the V2X communication, improves the control of CAVs and alleviates the computational load of the central control at the same time. In this case, the model predictive control (MPC), which is effective in solving multi-variable constrained control problems, is widely used. The combination of the distributed model predictive control (DMPC) can effectively improve the systematic stability and optimality (Venkat et al., 2005). For example, Goulet and Ayalew (2021) designed a DMPC framework to coordinately control CAVs in a traffic flow and guide HVs for an improved driving maneuver. The result shows an increasing fuel reduction with a higher MPR level, and the simulation proves that a 2D motion control is more effective than a 1D control on multi-lane roads. Similarly, the benefit of the lane-changing control is also illustrated in Jiang et al. (2022) and Shang et al. (2024).

6.3.2 Mitigating Network-Wide Environmental Impact from Routing Control

In addition to driving behavior control, routing algorithms help vehicles get rid of congested road links, leading to less congestion and lower emissions. Similar to eco-driving, eco-routing, which takes fuel-saving and emission reduction as parts of the optimization objectives, can be an effective method to mitigate the environmental impact of road traffic compared to the algorithms that only take traveling time cost as the only target (Aziz & Ukkusuri, 2012; Bektaş & Laporte, 2011; Guo et al., 2013).

Because of the potential side effect on traffic delay when taking the environmental impact as the only goal, studies have tried multi-objective optimization to simultaneously improve traffic throughput and emission reduction (Djavadian et al., 2020; Guo et al., 2015). Because of the higher robustness of the routing result compared to the central control, the distributed control gains popularity for eco-routing (Djavadian et al., 2020; Djavadian & Farooq, 2018). In this case, the communication between vehicles and surrounding facilities ensures the functionality of the routing control. For example, Elbery et al. (2016) found that if vehicles update their fuel consumption in real time, the optimized routing result can be elevated compared to the vehicles that are only able to send their information at street ends.

6.4 SUMMARY

This chapter introduces the method to estimate road traffic energy consumption and emissions, evaluation of mixed traffic through empirical data and simulation, environmental impact of the CAV introduction, and approaches for mitigating the side effects of CAVs from microscopic and macroscopic perspectives. Generally, the introduction of CAVs can alleviate energy usage and roadside air pollutants, but it highly depends on the pollutant type, control algorithm, and MPR of CAVs. The current literature highlights the importance of communication technologies that provide precise information about surrounding traffic, controlling strategies among which the distributed control should be prioritized, and controlling objectives that determine the weight of the environmental impact in the optimization.

REFERENCES

Almannaa, M. H., Chen, H., Rakha, H. A., Loulizi, A., & El-Shawarby, I. (2019). Field implementation and testing of an automated eco-cooperative adaptive cruise control system in the vicinity of signalized intersections. *Transportation Research Part D: Transport and Environment*, 67, 244–262.

Atkins. (2016). Impacts of connected and autonomous vehicles on traffic flow: evidence review. Technical Report, Department for Transport and Centre for Connected, and Autonomous Vehicles.

Aziz, H. M. A., & Ukkusuri, S. V. (2012). Integration of environmental objectives in a system optimal dynamic traffic assignment model. *Computer-Aided Civil and Infrastructure Engineering*, 27(7), 494–511.

Bektaş, T., & Laporte, G. (2011). The pollution-routing problem. *Transportation Research Part B: Methodological*, 45(8), 1232–1250.

Center for Environmental Research and Technology. (2017). Comprehensive Modal Emission Model (CMEM). www.cert.ucr.edu/cmem/

Chen, X., Zhu, M., Chen, K., Wang, P., Lu, H., Zhong, H., Han, X., Wang, X., & Wang, Y. (2023). Followнet: A comprehensive benchmark for car-following behavior modeling. *Scientific Data*, 10(1), 828.

Chen, Y., & Borken-Kleefeld, J. (2014). Real-driving emissions from cars and light commercial vehicles–Results from 13 years remote sensing at Zurich/CH. *Atmospheric Environment*, 88, 157–164.

Coelho, M. C., Fontes, T., Bandeira, J. M., Pereira, S. R., Tchepel, O., Dias, D., Sá, E., Amorim, J. H., & Borrego, C. (2014). Assessment of potential improvements on regional air quality modelling related with implementation of a detailed methodology for traffic emission estimation. *Science of the Total Environment*, 470, 127–137.

Colberg, C. A., Tona, B., Stahel, W. A., Meier, M., & Staehelin, J. (2005). Comparison of a road traffic emission model (HBEFA) with emissions derived from measurements in the Gubrist road tunnel, Switzerland. *Atmospheric Environment*, 39(26), 4703–4714.

de Eicker, M. O., Zah, R., Triviño, R., & Hurni, H. (2008). Spatial accuracy of a simplified disaggregation method for traffic emissions applied in seven mid-sized Chilean cities. *Atmospheric Environment*, 42(7), 1491–1502.

Di Pace, R., Fiori, C., Storani, F., de Luca, S., Liberto, C., & Valenti, G. (2022). Unified network tRaffic management frAmework for fully conNected and electric vehicles energy cOnsumption optimization (URANO). *Transportation Research Part C: Emerging Technologies*, 144, 103860.

Djavadian, S., & Farooq, B. (2018). Distributed dynamic routing using network of intelligent intersections. *ITS Canada ACGM*, 2018, 22.

Djavadian, S., Tu, R., Farooq, B., & Hatzopoulou, M. (2020). Multi-objective eco-routing for dynamic control of connected & automated vehicles. *Transportation Research Part D: Transport and Environment*, 87, 102513.

Elbery, A., Rakha, H., ElNainay, M. Y., Drira, W., & Filali, F. (2016). Eco-routing: An ant colony based approach. *International Conference on Vehicle Technology and Intelligent Transport Systems*, 2, 31–38.

Fameli, K. M., & Assimakopoulos, V. D. (2015). Development of a road transport emission inventory for Greece and the Greater Athens Area: Effects of important parameters. *Science of the Total Environment*, 505, 770–786.

Frey, H. C., Zhang, K., & Rouphail, N. M. (2010). Vehicle-specific emissions modeling based upon on-road measurements. *Environmental Science & Technology*, 44(9), 3594–3600.

Garcia-Castro, A., Monzon, A., Valdes, C., & Romana, M. (2017). Modeling different penetration rates of eco-driving in urban areas: Impacts on traffic flow and emissions. *International Journal of Sustainable Transportation*, 11(4), 282–294.

Goulet, N., & Ayalew, B. (2021). Distributed maneuver planning with connected and automated vehicles for boosting traffic efficiency. *IEEE Transactions on Intelligent Transportation Systems*, 23(8), 10887–10901.

Guo, C., Yang, B., Andersen, O., Jensen, C. S., & Torp, K. (2015). Ecosky: Reducing vehicular environmental impact through eco-routing. In *2015 IEEE 31st International Conference on Data Engineering* (pp. 1412–1415). IEEE.

Guo, L., Huang, S., & Sadek, A. W. (2013). An evaluation of environmental benefits of time-dependent green routing in the greater Buffalo–Niagara region. *Journal of Intelligent Transportation Systems*, 17(1), 18–30.

Hu, X., Zheng, Z., Chen, D., & Sun, J. (2023). Autonomous vehicle's impact on traffic: empirical evidence from Waymo Open Dataset and implications from modelling. *IEEE Transactions on Intelligent Transportation Systems*, 24(6), 6711–6724.

Hu, X., Zheng, Z., Chen, D., Zhang, X., & Sun, J. (2022). Processing, assessing, and enhancing the Waymo autonomous vehicle open dataset for driving behavior research. *Transportation Research Part C: Emerging Technologies*, 134, 103490.

Hu, X., Zheng, Z., Zhang, X., Chen, D., & Sun, J. (2021). Vehicle trajectory data processed from the Waymo open dataset. Technical Report, Department of Traffic Engineering, Tongji University, Shanghai, China.

Jiang, Z., Yu, D., Luan, S., Zhou, H., & Meng, F. (2022). Integrating traffic signal optimization with vehicle microscopic control to reduce energy consumption in a connected and automated vehicles environment. *Journal of Cleaner Production*, 371, 133694.

Kawsar, S., Biswas, S., Noor, M., & Mamun, M. S. (2024). Investigating the applicability of COPERT 5.5 emission software in Bangladesh and developing countrywide vehicular emission inventories. *Environmental Science: Atmospheres*, 4(1), 57–72.

Kopelias, P., Demiridi, E., Vogiatzis, K., Skabardonis, A., & Zafiropoulou, V. (2020). Connected & autonomous vehicles–Environmental impacts – a review. *Science of the Total Environment*, 712, 135237.

Li, D., Zhu, F., Wu, J., Wong, Y. D., & Chen, T. (2024). Managing mixed traffic at signalized intersections: An adaptive signal control and CAV coordination system based on deep reinforcement learning. *Expert Systems with Applications*, 238, 121959.

Li, G., Jiao, Y., Knoop, V. L., Calvert, S. C., & van Lint, J. W. C. (2023). Large Car-following Data based on Lyft level-5 Open Dataset: Following autonomous vehicles vs. human-driven vehicles. *ArXiv Preprint ArXiv:2305.18921*.

Lin, Y., Tiwari, A., Fabien, B., Ban, X., & Devasia, S. (2023). Increasing traffic capacity of mixed traffic at signalized traffic intersections using delayed self-reinforcement. *Transportation Research Part C: Emerging Technologies*, 157, 104403.

Luo, J., Barth, M. J., & Boriboonsomsin, K. (2018). Vehicle routing to mitigate human exposure to traffic-related air pollutants. In *2018 21st International Conference on Intelligent Transportation Systems (ITSC)* (pp. 2765–2770). IEEE.

Mahdinia, I., Mohammadnazar, A., Arvin, R., & Khattak, A. J. (2021). Integration of automated vehicles in mixed traffic: Evaluating changes in performance of following human-driven vehicles. *Accident Analysis & Prevention*, 152, 106006.

Pan, Y., Wu, Y., Xu, L., Xia, C., & Olson, D. L. (2024). The impacts of connected autonomous vehicles on mixed traffic flow: A comprehensive review. *Physica A: Statistical Mechanics and Its Applications*, 129454.

Progiou, A. G., & Ziomas, I. C. (2011). Road traffic emissions impact on air quality of the Greater Athens Area based on a 20 year emissions inventory. *Science of the Total Environment*, 410, 1–7.

Qin, Y., Wang, H., He, Z., and Ran, B. (2020a). Fuel consumption analysis of automated driving traffic flow based on vehicle specific power. *Journal of Transportation Systems Engineering and Information Technology*, 20(1), 91.

Qin, Y., Wang, H., He, Z., and Ran, B. (2020b). Stability analysis of traffic flow for automated vehicles. *Journal of Chongqing Jiaotong University of Science*, 39(12):20–25.

Qu, X., Zhong, L., Zeng, Z., Tu, H., & Li, X. (2022). Automation and connectivity of electric vehicles: Energy boon or bane? *Cell Reports Physical Science*, 3(8).

Rakha, H., Ahn, K., & Trani, A. (2004). Development of VT-Micro model for estimating hot stabilized light duty vehicle and truck emissions. *Transportation Research Part D: Transport and Environment*, 9(1), 49–74.

Robinette, D., Kostreva, E., Krisztian, A., Lackey, A., Morgan, C., Orlando, J., & Rama, N. (2019). PHEV real world driving cycle energy and fuel and consumption reduction potential for connected and automated vehicles. *Society of Automotive Engineers Technical Paper Series*, 1.

Shang, Y., Zhu, F., Jiang, R., Li, X., & Wang, S. (2024). Trajectory planning at a signalized road section in a mixed traffic environment considering lane-changing of CAVs and stochasticity of HDVs. *Transportation Research Part C: Emerging Technologies*, 158, 104441.

Stogios, C. (2018). *Investigating the Effects of Automated Vehicle Driving Operations on Road Emissions and Traffic Performance*. University of Toronto (Canada).

Tanvir, S. (2018). *Modeling and Simulation of Driving Activity from an Energy Use-Emissions Perspective*. North Carolina State University.

Tu, H., Zhao, L., Tu, R., & Li, H. (2024). The energy-saving effect of early-stage autonomous vehicles: A case study and recommendations in a metropolitan area. *Energy*, 297, 131274.

Tu, R., Du, J., Rakha, H. A., & Yang, H. (2017). *System-Wide Impacts of Arterial and Freeway Eco-Cooperative Adaptive Cruise Control.*

U.S. Environmental Protection Agency. (2023). Overview of EPA's MOtor Vehicle Emission Simulator (MOVES4). https://nepis.epa.gov/Exe/ZyPDF.cgi?Dockey=P10186IV.pdf

Venkat, A. N., Rawlings, J. B., & Wright, S. J. (2005). Stability and optimality of distributed model predictive control. In *Proceedings of the 44th IEEE Conference on Decision and Control* (pp. 6680–6685). IEEE.

Xu, H., Tu, R., Li, T., & Chen, H. (2023). Interpretable bus energy consumption model with minimal input variables considering powertrain types. *Transportation Research Part D: Transport and Environment*, 119, 103742.

Yao, Z., Wang, Y., Liu, B., Zhao, B., & Jiang, Y. (2021). Fuel consumption and transportation emissions evaluation of mixed traffic flow with connected automated vehicles and human-driven vehicles on expressway. *Energy*, 230, 120766.

Zhang, Y., Chen, X., Zhao, Y., Ma, J., & Yu, L. (2022). A methodology for measuring the environmental effect of autonomous bus considering platooning. *Transportation Research Part D: Transport and Environment*, 107, 103300.

7 Case Study

Driver Performance Difference and Model Difference Evaluation on Different Car-Following Scenarios under Mixed Traffic Flow

Yang Liu, Dengbo He, Xiaoyu Zhou, and Jiahao Zhang

7.1 INTRODUCTION

The high-level, SAE Level 4 or higher (SAE OnRoad Automated Vehicle Standards Committee, 2018) autonomous vehicles (AVs) running on the road have become a reality in many cities around the world. However, before the AVs saturate the market or designated lanes are built for AVs, the AVs will have to share the road with human-driven vehicles (HDVs), leading to mixed traffic. At the same time, it has been widely acknowledged that human operators may adapt to the technologies and adopt different strategies to work with them (Janssen et al., 2019; Kaber et al., 2009). Hence, to better model and manage the mixed traffic, we must model the human drivers' behaviors appropriately. Given that rear-end crashes dominate AV-involved crashes (Liu et al., 2021), this case study will mainly focus on the car-following behaviors of human drivers in mixed traffic.

So far, most of the studies on human drivers' behavior changes in mixed traffic were conducted via traffic/numerical simulations (Zheng et al., 2020) or field experiments (Mahdinia et al., 2021). A few studies have been conducted to model human drivers' behavior changes based on on-road data, due to the lack of real-world empirical data as a result of the low AV market penetration rate at this stage. This case study will demonstrate the application of real-world public AV driving datasets to model human drivers' car-following behaviors in mixed traffic and analyze the differences in human drivers' car-following behavior when the leading vehicle is AV versus HDVs.

The driving models used in this case include:

- Traditional car-following models, including the Intelligent Driver Model (IDM), Gazis–Herman–Rothery (GHR) car-following model (Al-Jameel, 2009), Gipps car-following model (Rakha et al., 2007), Wiedemann car-following model

DOI: 10.1201/9781032674667-7

(Abbas & Medina, n.d.), General Force (GF) model, and the Full Velocity Difference (FVD) model (Gong et al., 2008);
- Deep neural network models, including Deep Neural Network (DNN) and Long Short-Term Memory (LSTM);
- Car-following models based on deep reinforcement learning (DRL), including Proximal Policy Optimization (PPO) and Generative Adversarial Imitation Learning (GAIL).

All models were trained and calibrated based on the Waymo open dataset. The car-following model code in this case study is open to download at GitHub: https://github.com/lovezoyaxu/car-following-model-in-mixtraffic. If readers are interested in more car-following models, please refer to Chapter 3 for more information.

7.2 DATA DESCRIPTION

7.2.1 Waymo Open Dataset

The data used in this case is retrieved from the Waymo Open Dataset released by Waymo LLC. Waymo has been testing its AV fleet on public roads across various U.S. cities, accumulating over 32 million kilometers (Ettinger et al., 2021). The Waymo AVs are equipped with an array of sensors, including radars, lidars, and cameras, which can collect high-quality trajectory data of the vehicles and surrounding traffic at a frequency of 10 Hz. All the vehicles used in these tests can be categorized as SAE Level-4 or higher and have a uniform appearance (Di & Shi, 2021).

The Waymo Open Dataset is constituted of two parts: the perception and motion parts. In August 2019, Waymo released the perception part, including 1,000 20-second segments of well-synchronized and calibrated high-resolution lidar and camera data in urban and suburban environments (Sun et al., 2020). The motion portion of the dataset was released in March 2021, including 103,354 20-second long segments, totaling 574 hours of driving data. This data covers 1,750 kilometers of roadways across the U.S. Both perception and motion parts of the Waymo datasets were used in this case study. Interesting readers can find more information about the Waymo Open Dataset from the link: https://waymo.com/open/.

7.2.2 Car-Following Periods Extraction

In this case study, considering the sample size, the timespan of each segment, and the sensor detection range, the following criteria were adopted to extract qualified car-following events:

- The leading and following vehicles should be in the same lane on a straight segment;
- The lane changes and turns by either the leading or following vehicle were excluded;
- The space between the leading and following vehicles should be less than 85 m;
- The following vehicle's velocity should be greater than 1 m/s;
- The duration of the car-following event should be at least 10 seconds.

In this case, 171 HDV-following-AV events and 523 HDV-following-HDV events on highways were extracted from the Waymo Open Dataset. In the following analysis, 80% of the events were randomly selected as the training set of the model, and the remaining 20% were used for testing. This dataset is publicly available on GitHub: https://github.com/lovezoyaxu/car-following-model-in-mixtraffic.

7.3 COMPARISON BETWEEN HDV-FOLLOWING-HDV AND HDV-FOLLOWING-AV EVENTS

This section will reveal the difference in driver performance between HDV-following-HDV and HDV-following-AV events. In this case study, we mainly focus on two metrics that can quantify traffic stability and safety, that is, driving volatility, which reflects the comfort and fuel economy of the vehicle through the acceleration and deceleration fluctuations of the vehicle (Chai et al., 2022; Rios-Torres et al., 2019) and time headway, which can quantify the safety in car following events (Naito & Nagatani, 2012).

7.3.1 Driving Volatility

Driving aggression can be represented by driving volatility, which can be calculated from vehicle speed, acceleration, and deceleration data. Previous research has shown that higher driving volatility is associated with higher driver instability, leading to higher crash risk, more energy consumption, and more greenhouse gas emissions (Kamrani et al., 2018). Driving volatility can be quantified by the standard deviation of the vehicle speed, as shown in equation 7.1.

$$Std = \sqrt{\frac{\sum_{i=1}^{n}\left(v_i - \overline{v}\right)^2}{n-1}} \tag{7.1}$$

where v_i is the velocity value of observation, i, $\overline{v}$ is the mean of observations, and n is the sample size.

Figure 7.1 shows the density of the driving volatility when the HDV followed HDV versus AVs. Here, the density is calculated as the number of events in a volatility value that fell into a specific range out of all car-following events in that category (i.e., either HDV-following-HDV or HDV-following-AV). In general, when a human driver follows HDVs, the distribution of driving volatility is mostly concentrated in the range of 1.2–1.6 m/s, while when following AVs, the distribution of driving volatility is mostly concentrated in the range of 1.5–2 m/s. It is likely that drivers have different perceptions of AVs and thus exhibit different behaviors when following AVs (Mahdinia et al., 2021).

7.3.2 Driving Safety

Time headway is defined as the time difference between two consecutive vehicles passing through the same point. Time headway is calculated as the space headway

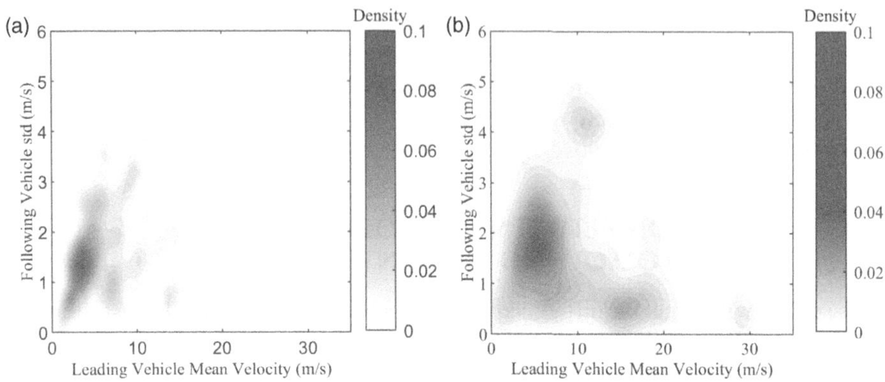

FIGURE 7.1 Velocity standard deviation distributions of following vehicles (a) HDV-following-HDV; (b) HDV-following-AV.

between the leading and following vehicles divided by the following vehicle's velocity, as shown in equation 7.2.

$$h_w(t) = \frac{X_L(t) - X_F(t)}{V_F(t)} \tag{7.2}$$

where $X_L(t)$ and $X_F(t)$ are the positions of the leading vehicle and following vehicle at time t, and $V_F(t)$ is the velocity of the following vehicle at time t.

The downward tendency can be observed in both subplots of Figure 7.2, suggesting that time headways decrease with the increments of average leading vehicle velocity. Again, it can be observed that when following AVs, the behaviors of HDVs are more diverse compared to those of HDVs.

7.4 CAR-FOLLOWING MODEL INTRODUCTION

In the car-following event, the state of the vehicle is defined as (1) the speed of the self-vehicle $V_n(t)$, (2) the relative speed between the self-vehicle and the lead vehicle $\Delta V_{n-1,n}(t)$, and (3) the distance between the self-vehicle and the lead vehicle $S_{n-1,n}(t)$. According to Newton's theory, the relationship between them is shown in the following three equations. The action that the following vehicle needs to take at time step t is the longitudinal acceleration, denoted as $a_n(t)$.

$$V_n(t+1) = V_n(t) + \Delta T \times a_n(t) \tag{7.3}$$

$$\Delta V_{n-1,n}(t+1) = V_n(t+1) - V_{n-1}(t+1) \tag{7.4}$$

$$S_{n-1,n}(t+1) = S_{n-1,n}(t) - \Delta T \times \frac{\Delta V_{n-1,n}(t) + \Delta V_{n-1,n}(t+1)}{2} \tag{7.5}$$

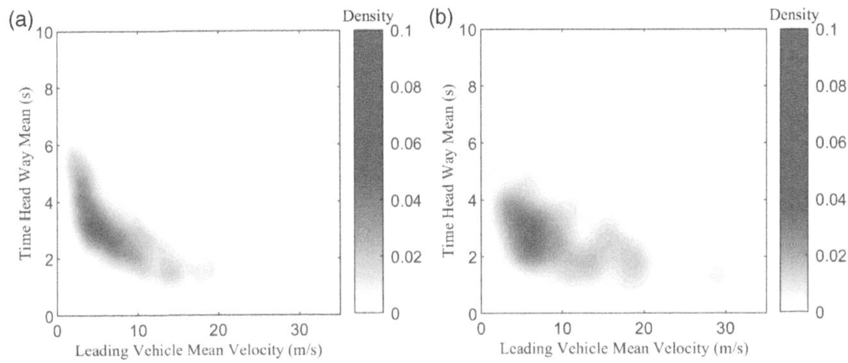

FIGURE 7.2 Time headway distributions (a) HDV-following-HDV; (b) HDV-following-AV.

where ΔT is the simulation time interval, which is the data collection interval and was set to be 0.1 s in this study, and $V_{n-1}(t+1)$ is the speed of the leading vehicle, which is set to be known over time.

The training process of one car-following event is set as an episode in this study. When a traffic crash happens (i.e., $S_{n-1,n}(t+1) \leq 0$) or the simulation reaches the maximum time step (i.e., the length of an event), all states will be re-initialized using the next car-following event data. Next, we will introduce several commonly used car-following models and illustrate how they can be applied to mixed traffic research.

7.4.1 Intelligent Driver Model

Microscopic car-following models have long been used to simulate individual driver behavior in the presence of preceding vehicles, of which the IDM is one of the most commonly used (Derbel et al., 2013):

$$a_n(t) = a_{max}^{(n)} \left(1 - \left(\frac{V_n(t)}{\widetilde{V}_n(t)}\right)^{\beta} - \left(\frac{\widetilde{S}_n(t)}{S_n(t)}\right)^2\right) \tag{7.6}$$

$$\widetilde{S}_n(t) = S_{jam}^{(n)} + max\left(0, V_n(t)\widetilde{T}_n(t) + \frac{V_n(t)\Delta V_n(t)}{2\sqrt{a_{max}^{(n)} a_{comfort}^{(n)}}}\right) \tag{7.7}$$

In the above equations, $a_{max}^{(n)}$ represent the maximum acceleration/deceleration of the following vehicle, $\widetilde{V}_n(t)$ is the desired speed, $S_{jam}^{(n)}$ is the minimum spacing at a standstill, $\widetilde{T}_n(t)$ is the desired time headway, $a_{comfort}^{(n)}$ is the comfortable deceleration, and β is the parameter that controls the degree of nonlinearity of acceleration behavior when the vehicle speed approaches the desired speed. We set it as 4 in this research by inserting the known state of the car-following trajectory into the above equations, we can obtain the desired action $a_n(t)$.

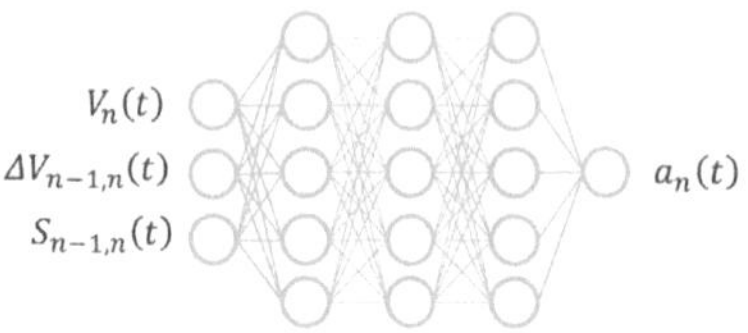

FIGURE 7.3 The structure of DNN.

7.4.2 Deep Neural Network

DNN (Canziani et al., 2016) is a machine learning model composed of multiple neural network layers. It typically consists of an input layer, multiple hidden layers, and an output layer, with each hidden layer containing multiple neurons. Each neuron is connected to all neurons in the previous layer, transmitting information through connections with associated weights. The structure of DNN is shown in Figure 7.3.

The core idea of DNN is to perform tasks by learning patterns and features from data. During training, DNN automatically adjusts connection weights to minimize the error between predicted results and actual labels. This is usually achieved through optimization algorithms like gradient descent, where weights are adjusted based on the error gradient, gradually converging the model toward the optimal solution. By inserting the known state of the car-following trajectory into the DNN, we can get the desired action $a_n(t)$. In this case study, we used three layers in the neural networks.

7.4.3 Long Short-Term Memory

LSTM (Hochreiter, 1997) is utilized to model the dynamic relationships between vehicles and traffic flow characteristics, enabling intelligent car-following strategies. By learning and predicting time-series data such as relative speeds, distances, and accelerations between vehicles, LSTM provides a deeper understanding, aiding the autonomous driving system in accurately determining when to accelerate, decelerate, or maintain distance for safe and efficient car-following. The objective function of the LSTM model is described as follows:

$$C(W,B) = \frac{\left(S_{n-1,n}(t) - S_{n-1,n}^{obs}(t)\right)^2}{\left(S_{n-1,n}^{obs}(t)\right)^2} \tag{7.8}$$

where $S_{n-1,n}^{obs}(t)$ is the observed spacing at time step t and W and B represent the weights and biases in the LSTM model. The LSTM model gradually minimizes the objective function by backpropagating a small update through time in the direction of optimizing the weights and biases. In this case study, we used a one-layer LSTM that considers 10 historical steps in the data.

7.4.4 Proximal Policy Optimization

To explain the PPO (Wei et al., 2019), we first introduce the Actor-Critic algorithm (ATC) (Xu et al., 2021), given that the PPO algorithm is evolved based on the ATC algorithm. We then assume that the target strategy π_θ is a random strategy and is differentiable everywhere, where θ is the corresponding parameter. We can use a linear model or a neural network model to model such a strategy function, input a state, and output a probability distribution of an action. Our goal is to find an optimal strategy and maximize the expected return of that strategy in the environment. We define the objective function of strategy learning as

$$J(\theta) = E_{s_0}\left[V^{\pi_\theta}(S_0)\right] \tag{7.9}$$

where S_0 represents the initial state. Now that we have the objective function, we take the derivative of the objective function with respect to the strategy, and once we get the derivative, we can use the gradient ascent method to maximize this objective function and get the optimal strategy.

$$\nabla_\theta J(\theta) = E_{\pi_\theta}\left[\sum_{t=0}^{T}\psi_t \nabla_\theta log_{\pi_\theta}(a_t \mid s_t)\right] \tag{7.10}$$

ATC algorithm is simple and intuitive, but it can be unstable during training in practical applications. This is because when the strategy network gradient updates the parameters, it is likely that the strategy will suddenly deteriorate significantly due to the long step size, thus affecting the training effect. To solve this problem, researchers added trust regions to ensure that each policy gradient update resulted in a performance improvement. Mathematically, we try to use the current θ to find a better parameter θ', so that $J(\theta') \geq J(\theta)$.

$$J(\theta') - J(\theta) = \frac{1}{1-\gamma} E_{s\sim v^{\pi_{\theta'}}} E_{a\sim\pi_{\theta'}(\cdot|s)}\left[A^{\pi_\theta}(s,a)\right] \tag{7.11}$$

The above equation is complicated to solve. Therefore, the state access distribution is processed accordingly. In particular, we ignore the change in the state access distribution between the two policies and directly adopt the state distribution of the old policy $\pi_{\theta'}$. The improved equation is as follows:

$$L_\theta(\theta') = J(\theta) + \frac{1}{1-\gamma} E_{s\sim v^{\pi_\theta}} E_{a\sim\pi_\theta(\cdot|s)}\left[\frac{\pi_{\theta'}(a \mid s)}{\pi_\theta(a \mid s)} A^{\pi_\theta}(s,a)\right] \tag{7.12}$$

The objective function of the above equation is restricted to ensure that the gap between the new and old parameters is not too large, and the resulting equation is the PPO update strategy.

$$argmax_{\theta} E_{s \sim v^{\pi_\theta}} E_{a \sim \pi_\theta(\cdot|s)} \left[min\left(\frac{\pi_\theta(a \mid s)}{\pi_k(a \mid s)} A^{\pi_\theta}(s,a), clip\left(\frac{\pi_\theta(a \mid s)}{\pi_k(a \mid s)}, 1-\varepsilon, 1+\varepsilon \right) A^{\pi_{\theta_k}}(s,a) \right) \right] \quad (7.13)$$

In this case, we used the reward function to facilitate human-like driving by minimizing the disparity between simulated and observed behavior values at each timestep. The reward equation is shown below:

$$r(t) = -\log\left(\left| \frac{S_{sim}(t) - S_{obs}(t)}{S_{obs}(t)} \right| \right) \quad (7.14)$$

where $S_{sim}(t)$ is the spacing at time t given by the algorithm, and $S_{obs}(t)$ is the real spacing. When using PPO, we put the state of the car-following trajectory into policy network π_θ and use equation 7.13 and equation 7.14 to train the policy network. The optimum policy network will give the desired $a_n(t)$ and its corresponding probability at each state.

7.4.5 Generative Adversarial Imitation Learning

The GAIL (Ho & Ermon, 2016) is based on the Generative Adversarial Network (GAN) (Creswell et al., 2018). GAN consists of two main components: a generator and a discriminator. These two networks compete with each other through adversarial learning to improve the performance of the generator model.

The goal of a generator is to generate realistic data samples from random noise, such as images, audio, or text. It takes a random vector or noise as input and attempts to map it into a space similar to the real data sample. In contrast, the discriminator is used to distinguish between fake samples generated by the generator and real data samples. It takes samples from the generator as well as samples from the real data set and tries to tell them apart.

The main difference between the GAN and the GAIL is the generator. The generator of GAIL consists of an agent that can generate actions based on state. A typical structure of GAIL is shown in Figure 7.4. In our case study, the generator uses the PPO algorithm improved by the policy gradient.

When using GAIL, firstly, an expert trajectory, which is the recorded real-world trajectory from the Waymo dataset, was randomly selected, and the state of the trajectory was initialized. Then, the initial state and leading vehicle data are input into the generator, and the PPO algorithm generates a trajectory based on this data. The discriminator optimizes itself by discriminating between the generated trajectory and the expert trajectory. The pseudo-reward generated by the discriminator will help optimize the PPO algorithm. The loss function of the discriminator is as follows:

$$L(\psi) = -E_{\rho\pi}\left[logD_\psi(s,a) \right] + E_{\rho E}\left[log\left(1 - D_\psi(s,a)\right) \right] \quad (7.15)$$

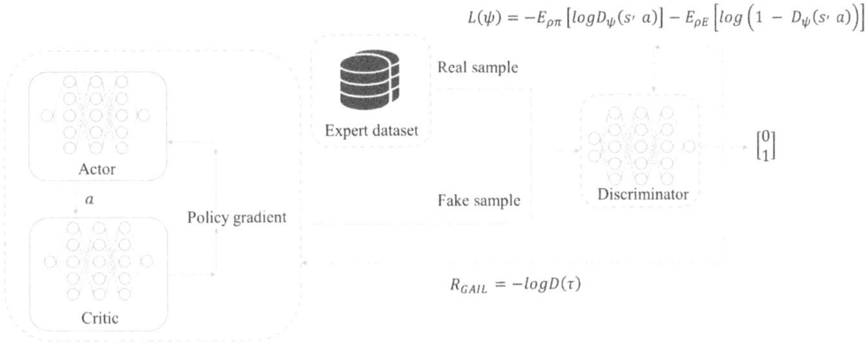

FIGURE 7.4 The structure of GAIL.

In equation 7.15, D_{ψ} represent discriminator network, $E_{\rho E}$ and $E_{\rho\pi}$ represent the expectations for state-action pairs of expert policy and generator policy, respectively. The discriminator tries to push the output of the expert data close to 0 and the output of the mimic policy close to 1. Thus, when minimizing the loss function, the discriminator attempts to distinguish between data generated by expert strategies and data generated by generator strategies. At the same time, the generator tries to generate data that "cheats" the discriminator, thus approximating the expert strategy. The equation that the discriminator provides for the generator of pseudo-rewards is

$$R_{GAIL} = -logD(s, a) \tag{7.16}$$

When the discriminator believes that the trajectory generated by PPO is close to the expert trajectory, the pseudo-reward will become large, encouraging the PPO generation to generate the trajectory in this way. Otherwise, the pseudo-reward will become small, inhibiting the PPO from generating the trajectory in such a way.

The performance of the GAIL can be expressed by equation 7.14. It should be noted that this reward is different from the pseudo-reward generated by the discriminator. In the GAIL algorithm, PPO is only updated by the pseudo-reward of equation 7.16. Equation 7.14 only reflects the quality of GAIL training results. However, in the PPO algorithm, equation 7.14 reward is used to update the PPO algorithm. You may compare the open-source code provided in this case study to distinguish the difference between these two algorithms.

Figure 7.5 shows how the performance of the GAIL model changes with training time for different car following types (i.e., HDV-following-HDV versus HDV-following-AV). In general, the model becomes stable after 800 episodes.

7.5 CAR-FOLLOWING MODEL PERFORMANCE

To give readers an intuitive sense of the differences among different models, this section compares the spacing between the following vehicle and the leading vehicle and the speed of the following vehicle velocity generated by different algorithms.

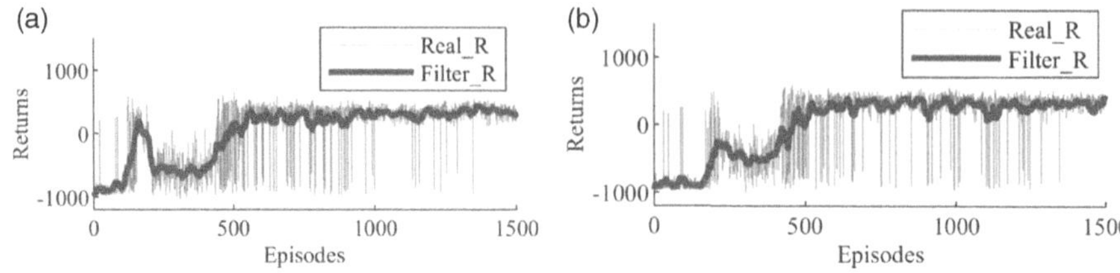

FIGURE 7.5 GAIL model training return change trend (a) HDV-following-HDV; (b) HDV-following-AV.

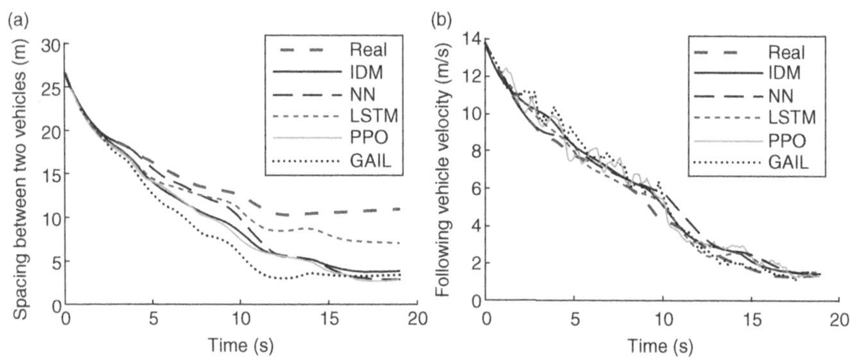

FIGURE 7.6 Car-following model performance in HDV-following-HDV scenarios.

Figure 7.6 shows the HDV-following-HDV result, and Figure 7.7 shows the HDV-following-AV results.

It can be observed that compared with DRL algorithms (i.e., PPO and GAIL), IDM and DNN car-following algorithms (i.e., DNN and LSTM) are smoother. DRLs led to frequent fluctuations in speed. In addition, the error between the real trajectory and the LSTM algorithm is the smallest among all algorithms in terms of spacing and velocity.

Tables 7.1 and 7.2 provide quantitative results regarding the performance of different cars following models on the test dataset. In the tables, MSE means Squared Error. MSE (spacing) and MSE (velocity) are the averages of the square of the spacing and velocity errors between each generated trajectory and the expert trajectory, respectively. The jerk metric evaluates the smoothness of vehicle acceleration or deceleration. It measures the rate of change of acceleration and reflects the comfort level experienced by the passengers of the ego vehicle. A lower jerk value indicates a smoother and more comfortable driving experience. TTC is Time-To-Collision. TTC is a commonly used surrogate safety measure, which was initially introduced in (Vogel, 2003) and defined as "*the time that remains until a crash between two vehicles would have occurred if the crash course and speed difference are maintained.*" as shown below:

$$TTC(t) = \frac{X_L(t) - X_F(t) - L}{V_F(t) - V_L(t)} \tag{7.17}$$

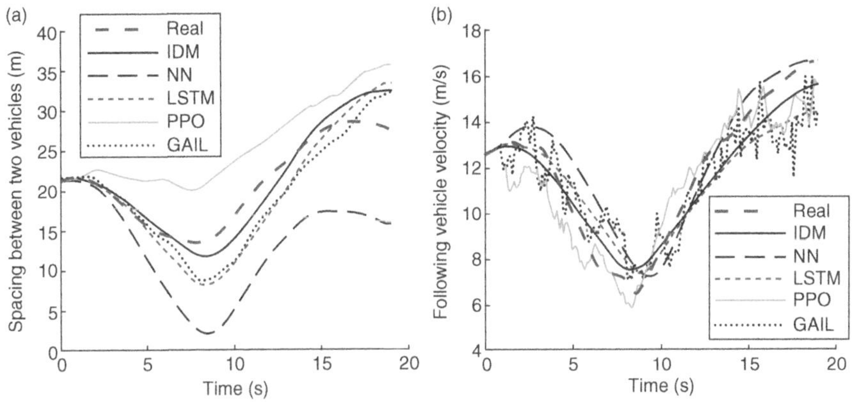

FIGURE 7.7 Car-following model performance in HDV-following-AV scenarios.

TABLE 7.1
Model performance comparison in HDV-following-HDV scenarios

Test standard \ Model	IDM	DNN	LSTM	PPO	GAIL
Test of MSE (spacing)	37.31	54.66	**26.76**	48.49	46.81
Test of MSE (velocity)	1.25	1.48	**0.99**	1.17	3.78
Jerk	0.20	0.28	**0.27**	8.91	19.81
TTC	**34.43**	14.35	5.26	23.39	17.12

TABLE 7.2
Model performance comparison in HDV-following-AV scenarios

Test standard \ Model	IDM	DNN	LSTM	PPO	GAIL
Test of MSE (spacing)	32.06	21.41	**20.47**	23.85	41.43
Test of MSE (velocity)	45.52	41.48	**38.75**	45.50	49.04
Jerk	12.31	**6.48**	9.71	7.73	20.76
TTC	114.94	116.56	119.66	124.52	**123.37**

where $X_L(t)$ and $X_F(t)$ are the positions of centers of mass of the leading and following vehicles at time t; L is the length of the vehicles; and $V_F(t)$ and $V_L(t)$ are the velocity of the leading and following vehicles, respectively, at time t.

The descriptive statistical results in Tables 7.1 and 7.2 are the average values without outliers. As can be seen from the two tables, the LSTM has the best performance when recovering the car-following trajectory. DRL algorithms, especially the GAIL algorithm, led to the largest jerk values, which can impair driving comfort and fuel economy. Compared with following HDVs, TTC is significantly improved when following AVs. This is because, in the Waymo dataset, the spacing between vehicles is larger when following AVs versus that when following HDVs.

7.6 CONCLUSION

This case study analyzes the performance differences of human drivers when following different types of vehicles by extracting car-following events from the Waymo dataset. In addition, we demonstrated the application of five different algorithms in recovering car-following behaviors in mixed traffic. The results show that the LSTM algorithm has the best performance in terms of approaching the real trajectory, and DRL algorithms tend to lead to large jerk. The readers can refer to the sample codes published in GitHub for more details regarding the five algorithms. Readers can adjust the parameters in the sample codes to get better results.

REFERENCES

Al-Jameel, H. a. E. (2009). *Examining and improving the limitations of Gazis–Herman–Rothery car following model.* https://salford-repository.worktribe.com/output/1459924

Canziani, A., Paszke, A., & Culurciello, E. (2016). An analysis of deep neural network models for practical applications. *arXiv preprint*. arXiv:1605.07678.

Chai, C., Shi, X., Zhou, Z., Zeng, X., Yin, W., & Islam, Md. M. (2022). Driving style recognition based on naturalistic driving: Volatilities, decision-making, and safety performances. In A. Riener, M. Jeon, & I. Alvarez (Eds.), *User Experience Design in the Era of Automated Driving* (pp. 359–394). Springer International Publishing. https://doi.org/10.1007/978-3-030-77726-5_14

Creswell, A., White, T., Dumoulin, V., Arulkumaran, K., Sengupta, B., & Bharath, A. A. (2018). *Generative adversarial networks: An overview. IEEE Signal Processing Magazine*, 35(1), 53–65. https://doi.org/10.1109/MSP.2017.2765202

Derbel, O., Peter, T., Zebiri, H., Mourllion, B., & Basset, M. (2013). Modified Intelligent Driver Model for driver safety and traffic stability improvement. *IFAC Proceedings Volumes*, 46(21), 744–749. https://doi.org/10.3182/20130904-4-JP-2042.00132

Di, X., & Shi, R. (2021). A survey on autonomous vehicle control in the era of mixed-autonomy: From physics-based to AI-guided driving policy learning. *Transportation Research Part C: Emerging Technologies*, 125, 103008. https://doi.org/10.1016/j.trc.2021.103008

Ettinger, S., Cheng, S., Caine, B., Liu, C., Zhao, H., Pradhan, S., Chai, Y., Sapp, B., Qi, C. R., Zhou, Y., Yang, Z., Chouard, A., Sun, P., Ngiam, J., Vasudevan, V., McCauley, A., Shlens, J., & Anguelov, D. (2021). *Large scale interactive motion forecasting for autonomous driving: The Waymo Open Motion Dataset*. 9710–9719. https://openaccess.thecvf.com/content/ICCV2021/html/Ettinger_Large_Scale_Interactive_Motion_Forecasting_for_Autonomous_Driving_The_Waymo_ICCV_2021_paper.html

Gong, H., Liu, H., & Wang, B.-H. (2008). An asymmetric full velocity difference car-following model. *Physica A: Statistical Mechanics and Its Applications*, 387(11), 2595–2602. https://doi.org/10.1016/j.physa.2008.01.038

Ho, J., & Ermon, S. (2016). Generative adversarial imitation learning. *Advances in Neural Information Processing Systems*, 29. https://proceedings.neurips.cc/paper_files/paper/2016/hash/cc7e2b878868cbae992d1fb743995d8f-Abstract.html

Janssen, C. P., Donker, S. F., Brumby, D. P., & Kun, A. L. (2019). History and future of human–automation interaction. *International Journal of Human-Computer Studies*, 131, 99–107. https://doi.org/10.1016/j.ijhcs.2019.05.006

Hochreiter, S. (1997). *Long Short-term Memory. Neural Computation*. MIT-Press.

Kaber, D. B., Stoll, N., Thurow, K., Green, R. S., Kim, S.-H., & Mosaly, P. (2009). Human–automation interaction strategies and models for life science applications. *Human Factors and Ergonomics in Manufacturing & Service Industries*, 19(6), 601–621. https://doi.org/10.1002/hfm.20156

Kamrani, M., Arvin, R., & Khattak, A. J. (2018). Extracting useful information from connected vehicle data: An empirical study of driving volatility measures and crash frequency at intersections. *Transportation Research Record*, 2672(38), 290–301.

Liu, Q., Wang, X., Wu, X., Glaser, Y., & He, L. (2021). Crash comparison of autonomous and conventional vehicles using pre-crash scenario typology. *Accident Analysis & Prevention*, 159, 106281. https://doi.org/10.1016/j.aap.2021.106281

Mahdinia, I., Mohammadnazar, A., Arvin, R., & Khattak, A. J. (2021). Integration of automated vehicles in mixed traffic: Evaluating changes in performance of following human-driven vehicles. *Accident Analysis & Prevention*, 152, 106006. https://doi.org/10.1016/j.aap.2021.106006

Naito, Y., & Nagatani, T. (2012). Effect of headway and velocity on safety–collision transition induced by lane changing in traffic flow. *Physica A: Statistical Mechanics and Its Applications*, 391(4), 1626–1635. https://doi.org/10.1016/j.physa.2011.09.020

Rakha, H., Pecker, C. C., & Cybis, H. B. B. (2007). Calibration procedure for Gipps car-following model. *Transportation Research Record*, 1999(1), 115–127. https://doi.org/10.3141/1999-13

Rios-Torres, J., Liu, J., & Khattak, A. (2019). Fuel consumption for various driving styles in conventional and hybrid electric vehicles: Integrating driving cycle predictions with fuel consumption optimization. *International Journal of Sustainable Transportation*, 13(2), 123–137. https://doi.org/10.1080/15568318.2018.1445321

SAE OnRoad Automated Vehicle Standards Committee. (2018). *SAE levels of driving Automation™ refined for clarity and international audience*. www.sae.org/site/blog/sae-j3016-update

Sun, P., Kretzschmar, H., Dotiwalla, X., Chouard, A., Patnaik, V., Tsui, P., Guo, J., Zhou, Y., Chai, Y., Caine, B., Vasudevan, V., Han, W., Ngiam, J., Zhao, H., Timofeev, A., Ettinger, S., Krivokon, M., Gao, A., Joshi, A., … Anguelov, D. (2020). *Scalability in perception for autonomous driving: Waymo Open Dataset*. 2446–2454. https://openaccess.thecvf.com/content_CVPR_2020/html/Sun_Scalability_in_Perception_for_Autonomous_Driving_Waymo_Open_Dataset_CVPR_2020_paper.html

Vogel, K. (2003). A comparison of headway and time to collision as safety indicators. *Accident Analysis & Prevention*, 35(3), 427–433. https://doi.org/10.1016/S0001-4575(02)00022-2

Wei, H., Liu, X., Mashayekhy, L., & Decker, K. (2019). *Mixed-autonomy traffic control with proximal policy optimization*. In *2019 IEEE Vehicular Networking Conference (VNC)* (pp. 1–8), Los Angeles, California. https://doi.org/10.1109/VNC48660.2019.9062809

Xu, Z., Yang, D., Tang, J., Tang, Y., Yuan, T., Wang, Y., & Xue, G. (2021). An actor–critic-based transfer learning framework for experience-driven networking. *IEEE/ACM Transactions on Networking*, 29(1), 360–371. https://doi.org/10.1109/TNET.2020.3037231

Zheng, F., Liu, C., Liu, X., Jabari, S. E., & Lu, L. (2020). Analyzing the impact of automated vehicles on uncertainty and stability of the mixed traffic flow. *Transportation Research Part C: Emerging Technologies*, 112, 203–219. https://doi.org/10.1016/j.trc.2020.01.017

8 Case Study
Analysis of Mixed Traffic Flow

Jian Rong, Yi Wang, Chenjing Zhou, and Yue Chen

8.1 MICROSCOPIC PARAMETER CALIBRATION

This section introduces a novel parameter calibration method for VISSIM. Currently, the parameters that need to be calibrated include vehicle type, speed distribution, vehicle acceleration/deceleration distribution, as well as car-following and lane-changing driving behavior characteristic parameters (see Table 8.1).

8.1.1 Calibration Scheme

The calibration scheme in this chapter is mainly divided into four parts: selection of verification indicators and division of parameter calibration system, parameter sensitivity analysis, design of the objective function and genetic algorithm parameter calibration, and validation of the calibration model.

8.1.1.1 Step 1: Selection of Verification Indicators and Division of Parameter Calibration System

We extended the selection of verification indicators for parameter calibration and proposed an optimization approach based on multi-point curve calibration. Subsequently, we refined the research on the parameter calibration system and defined global parameters and local parameters separately. Global parameters were calibrated using engineering measurement methods, while local parameters were optimized and calibrated using genetic algorithms.

8.1.1.2 Step 2: Parameter Sensitivity Analysis

To reduce the variety of calibration objects and decrease the simulation workload, we combined the concept of variance analysis. By comparing the impact of simulation parameters and random seeds on verification indicators, sensitivity analysis was performed on local parameters to identify significant parameters and determine verification indicators.

DOI: 10.1201/9781032674667-8

TABLE 8.1
Summary of VISSIM simulation calibration parameters

Model parameters	Detail parameters	Range
Vehicle type	• Length (m)/width (m)	None
Speed distribution	• Expected speed distribution (km/h)	None
Vehicle acceleration/ deceleration distribution	• Maximum and expected acceleration distribution (m/s^2)	None
	• Maximum and expected deceleration distribution (m/s^2)	None
Car-following behavior module	• Maximum frontal distance(m) /OB_d	(150, 350)
	• Number of observed vehicles (count)/ N_pr	(1, 5)
	• Average parking gap (m)/ AX_average	(1, 4)
	• Additional part of safe distance (m)/BX_add	(1, 5)
	• Multiple part of safe distance (m)/BX_mult	(1, 6)
Lane change behavior module	• Maximum deceleration (m/s^2)/ACC_max	(−6, −2)
	• Acceptable deceleration (m/s^2)/ACC_ac	(−2, −0.5)
	• −1 m/s^2/distance (m)/D	(50, 250)
	• Waiting time before disappearance(s)/T_dis	(20, 60)
	• Minimum headway distance (m)/GAP_min	(0.5, 5)
	• Safe distance conversion coefficient/ABX	(0.1, 1)
	• Maximum deceleration of combined braking (m/s^2)/CO_ac	(−5, −1)

8.1.1.3 Step 3: Design of Objective Function and Genetic Algorithm Parameter Calibration

Based on multi-point distribution curves and incorporating the dynamic time warping (DTW) algorithm, we designed an objective function. Then, we utilized genetic algorithms to perform simulation local parameter calibration and obtained optimal local parameters through the process of "cluster recursion and classification averaging."

8.1.1.4 Step 4: Validation of the Calibration Model

Design different types of experimental schemes and select indicators such as the cumulative distance of the curves from the DTW algorithm and the *P*-value from the Kolmogorov–Smirnov (K-S) test to evaluate the effectiveness of the calibration model (see Figure 8.1).

8.1.2 Simulation Scene Configuration

The intersection of NanHeng Street and Caishikou Street in Beijing is selected as the research object, with 10 lanes in both north and south directions and 6 lanes in both east and west directions, each lane with a width of 3.5 m. The signal control conditions follow a three-phase timing plan with a cycle of 159 seconds. Based on the collected data, a simulation scene of the intersection of NanHeng Street and Caishikou Street in Beijing is constructed using VISSIM, as shown in Figure 8.2.

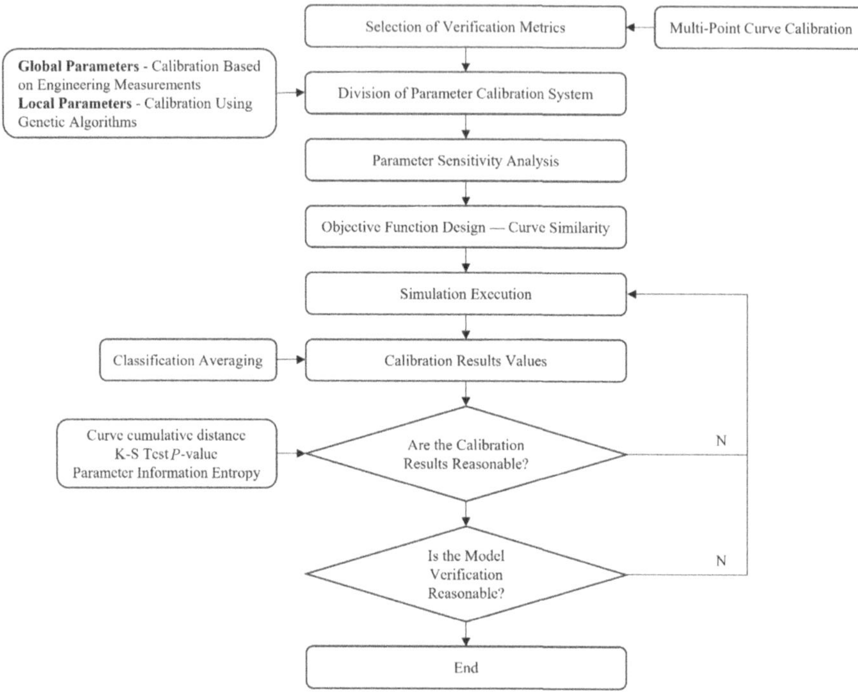

FIGURE 8.1 Microscopic traffic simulation model parameter calibration. Framework.

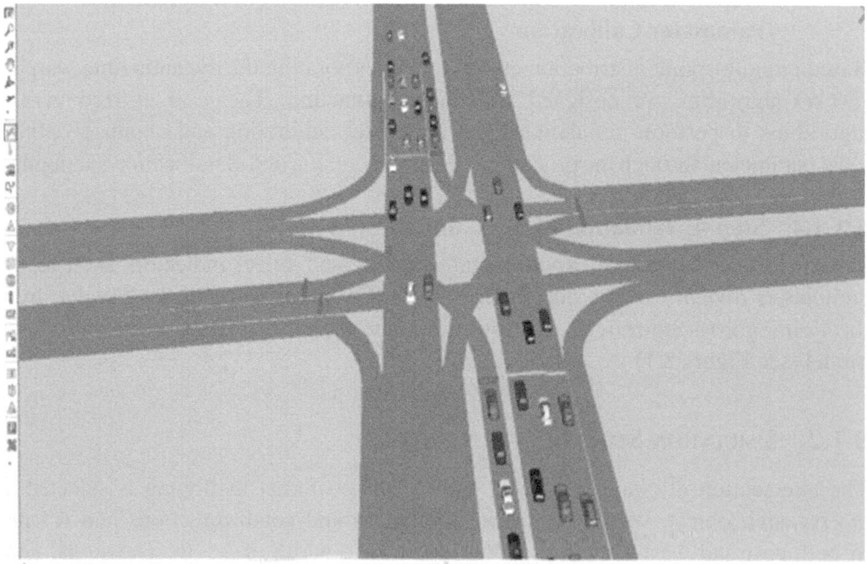

FIGURE 8.2 Simulation scene of the intersection of NanHeng Street and Caishikou Road.

8.1.3 Adopted Model Parameters and Validation Measure

Common indicators such as operating delay, travel time, queue length, discharge flow, and operating speed are selected as verification criteria. Simulation experiments are conducted to construct a sensitivity analysis matrix. Taking the average stopping distance in delay and car-following model parameters as an example, the average stopping distance is evenly divided into five parts within the range of values. The simulation duration is set to 3600 seconds, and random seeds correspond to five parts, that is, 20, 60, 100, 140, 180. In each simulation experiment, only one parameter or random seed changes while others remain constant. According to this principle, sensitivity analysis experiments were conducted for 12 types of simulation parameters to obtain simulation results. The specific application process of the sensitivity analysis method is as follows:

Constructing the Sensitivity Analysis Matrix: The columns represent simulation results with different random seed values under the same model parameter value, as shown in equation 8.1.

Model Parameter Sensitivity Analysis: According to equation 8.2, determine the standard deviation of each row and column step by step. On this basis, summarize and organize to calculate the mean of the standard deviations of the matrix's horizontal and vertical columns according to equations 8.3 and 8.4. Finally, determine whether the model parameter is sensitive to the calibration indicator based on equation 8.5.

$$X_{ij} = \begin{bmatrix} X_{11} & \cdots & X_{1m} \\ \vdots & \ddots & \vdots \\ X_{n1} & \cdots & X_{nm} \end{bmatrix} \tag{8.1}$$

$$STDEVA\left(X_i\right) = \sqrt{\frac{\sum_{i=1}^{n}\left(X_i - \bar{X}\right)^2}{n-1}} \tag{8.2}$$

$$I = \frac{\sum_{i=1}^{m} STDEVA\left(X_{1i}, X_{2i} \ldots X_{mi}\right)}{m} \tag{8.3}$$

$$K = \frac{\sum_{j=1}^{n} STDEVA\left(X_{j1}, X_{j2} \ldots X_{jm}\right)}{n} \tag{8.4}$$

$$P = \frac{I}{K} \tag{8.5}$$

In the equation:

X_{ij} represents the sensitivity analysis matrix of simulation experiments.

$i = 1, 2, 3 \ldots m$ indicates the levels of variation of simulation parameters.

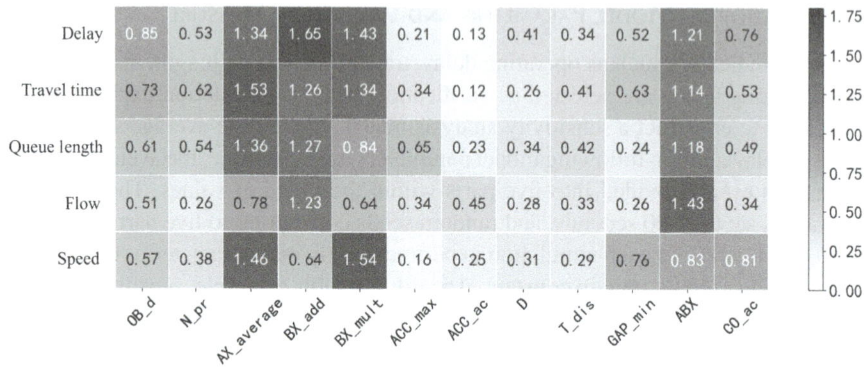

FIGURE 8.3 Results of parameter sensitivity analysis matrix.

$j = 1,2,3 \ldots n$ represents the levels of random seed selection.
$\text{STDEVA}(X_i)$ denotes the numerical standard deviation.
I stands for the mean standard deviation of simulation results for different parameter values.
K represents the mean standard deviation of simulation results for different random seeds.
P is the indicator for determining the results of the matrix. If the P-value is greater than 1, it is considered that the model parameter affects the calibration indicators; if the P-value is less than 1, it is considered that the simulation parameters do not affect the calibration indicators.

Organize all experimental results as shown in Figure 8.3. Taking delay and travel time as the calibration basis, there are four significant parameters with P-values greater than 1, namely average parking distance, additional part of the safety distance, multiple parts of the safety distance, and conversion coefficient of the safety distance, and likewise for others. Delay and travel time are most affected by model parameters. Considering the convenience of obtaining verification indicators, travel time is ultimately selected as the verification indicator, with other indicators as validation indicators. The parameters to be calibrated are determined to be the average stopping distance, additional part of the safety distance, multiple part of the safety distance, and conversion coefficient of the safety distance.

8.1.4 Multi-Point Curve Calibration

The investigation area is selected as a straight section 50 m behind the stop line, as shown in Figure 8.4(a). The specific data collection method involves timing from when a vehicle enters the investigation area, tracking vehicles entering the area, and counting the time and number of vehicles passing through the investigation area every 10 seconds. This data is then used to generate the measured travel time cumulative distribution curve, as shown in Figure 8.4(b). The mean travel time within

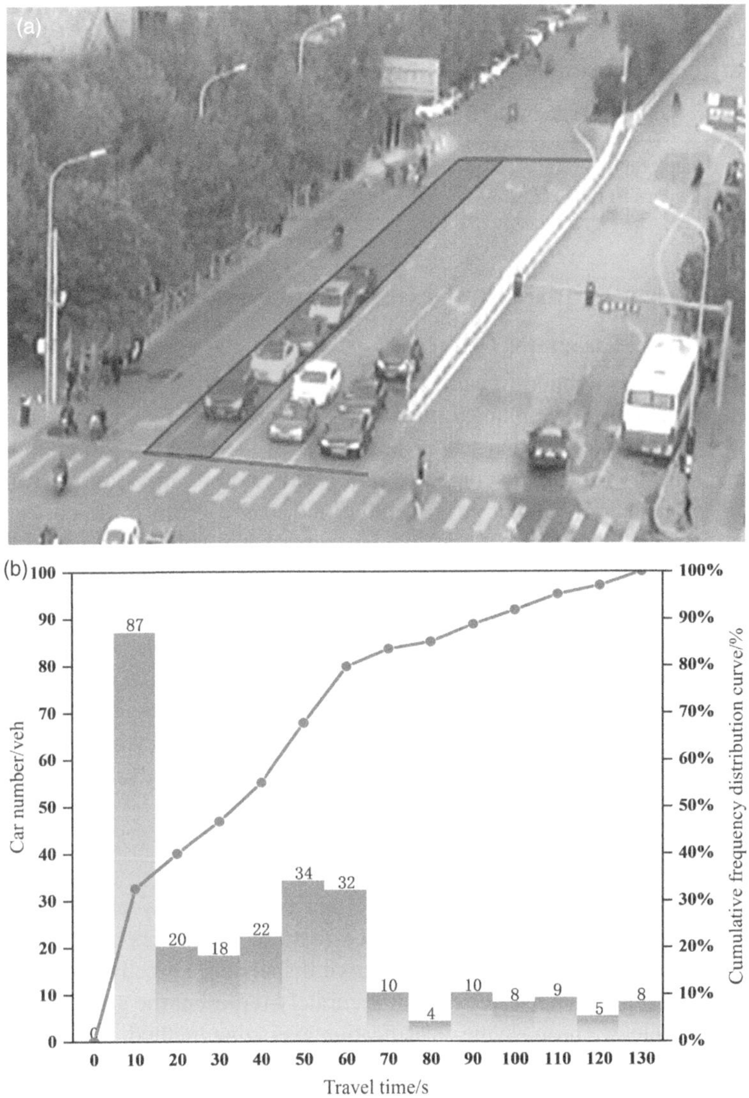

FIGURE 8.4 Measured travel time collection (a) Travel time collection data area; (b) cumulative distribution curve of travel time.

one hour from the measured data is used as the calibration basis for the single-point mean calibration method to carry out the model parameter calibration work. The DTW algorithm is employed to design the objective function for the multi-point curve calibration method, serving as the basis for conducting simulation parameter calibration.

TABLE 8.2
Comparison of maximum acceleration and deceleration values

Parameter	Parameter default value	Parameter calibrated value
Maximum acceleration	0–3.28	0–2.17
Maximum deceleration	−3.48–0	−2.25–0

8.1.5 Calibration through Parameter Categorization

8.1.5.1 Global Parameter Calibration

The V-BOX tool is applied to calibrate globally meaningful parameters, leveraging the measurement results of maximum acceleration, deceleration, expected acceleration, and deceleration from Lin et al. (2019) as the basis for calibrating global parameters in this paper. The comparison of results before and after the calibration of global parameters is shown in Table 8.2, and, based on this, the corresponding settings for global parameters in the simulation model are adjusted accordingly.

8.1.5.2 Local Parameter Calibration

Based on the sensitivity analysis matrix results in Figure 8.3, the local parameters to be calibrated are determined to be the safety distance conversion coefficient, safety distance additional part, safety distance multiplier, and average parking gap. Using the optimization experiment method for calibration indicators, Python is employed to call the VISSIM COM interface, and genetic algorithms are applied to calibrate local parameters.

8.1.5.3 Classification Averaging

The genetic algorithm terminates under certain conditions. By plotting the scatter plot of parameter distribution, it is observed that the parameters data that meet the termination condition of iteration have a large dispersion. Directly taking the mean as the calibrated parameter value cannot accurately represent the actual situation. To address this issue, we propose the classification averaging method, using information entropy as a quantifiable indicator of the parameter distribution state to guide the step-by-step determination of the calibration results.

Extract model parameter combinations that meet the termination condition of the genetic algorithm iteration as analysis objects. Normalize each model parameter to distribute them within the range of 0–1, as shown in Figure 8.5(a). On this basis, apply the elbow method and clustering method to cluster each parameter. Taking parameter AX_average as an example, apply the elbow method to determine the optimal number of clusters for parameter AX_average as 3, and then apply the clustering algorithm to cluster parameter AX_average into 3 classes, as shown in Figure 8.5(b). The clustering process for other parameters is the same as above.

Following the principle of "the smaller the entropy, the priority in parameter selection," the parameters were initially defined based on the clustering results of the system with the smallest information entropy. The mean value of the data with the

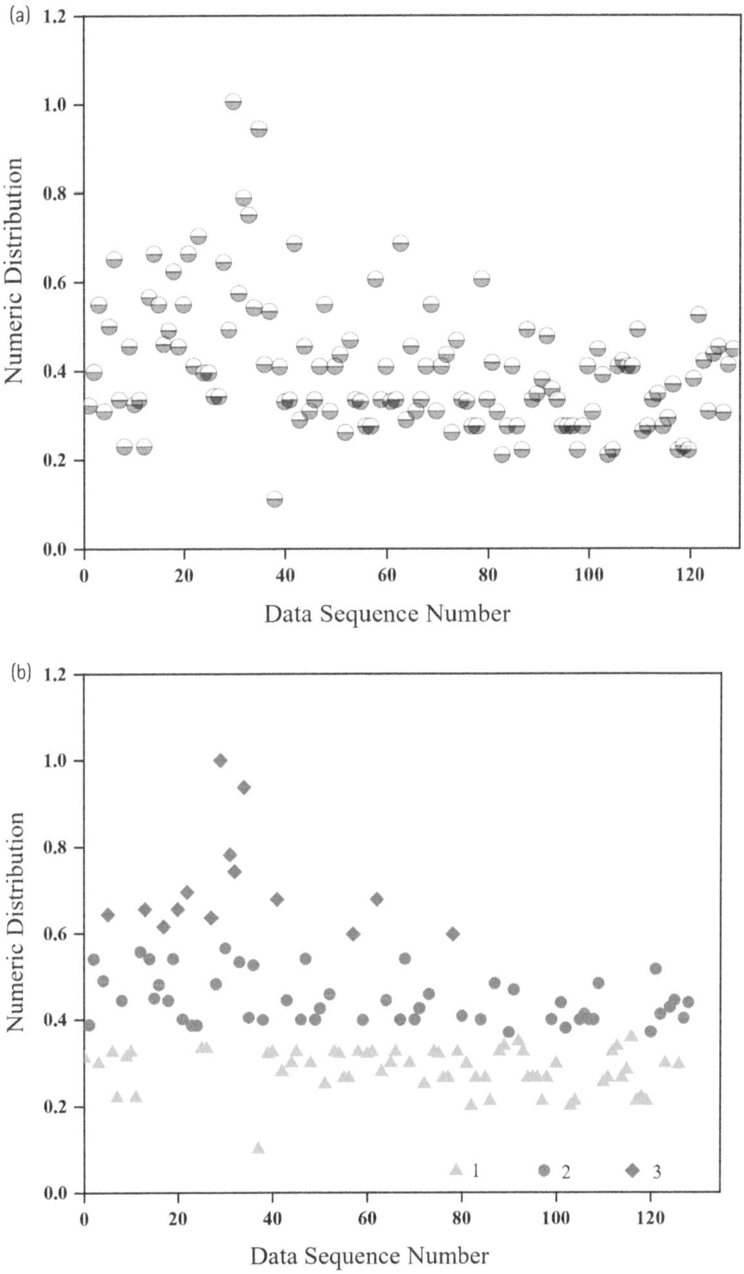

FIGURE 8.5 Cluster determination process of AX_average.

highest proportion in the corresponding subclass was taken as the calibration value for that parameter. Simultaneously, other parameter combinations in other subclasses were excluded. After determining the calibration value for one parameter, the information entropy of the remaining three parameters was recalculated. Following the above value determination rules and methods, each parameter value was determined individually.

8.1.6 Model Validation

Simulation calibration was carried out by designing eight experimental scenarios to select the optimal experimental setup. The experimental plans are as shown in Figure 8.6.

The final parameter calibration values for each experimental scheme are shown in Table 8.3.

This paper selects the intersection of Shilouzhuang Road and Liuxiang Road in Beijing to verify the portability of the optimal experimental model. The parameter calibration values of the eight experimental schemes mentioned above are applied to simulate the operation of the Shilouzhuang Road and Liuxiang Road intersection. The simulated output results for travel time and the cumulative distribution curve of measured travel time are shown in Figure 8.7.

Based on Figure 8.7 and Table 8.4, it is observed that for Scheme 1, the cumulative distance of the simulated output results from the calibration parameter values to the statistical results of actual travel times is minimal, and the p-value from the K-S test is maximal. This indicates that the simulated results of Scheme 1 better fit the cumulative distribution curve of actual travel times, demonstrating better portability of the calibrated parameter results. Therefore, it validates the rationality of the entire system

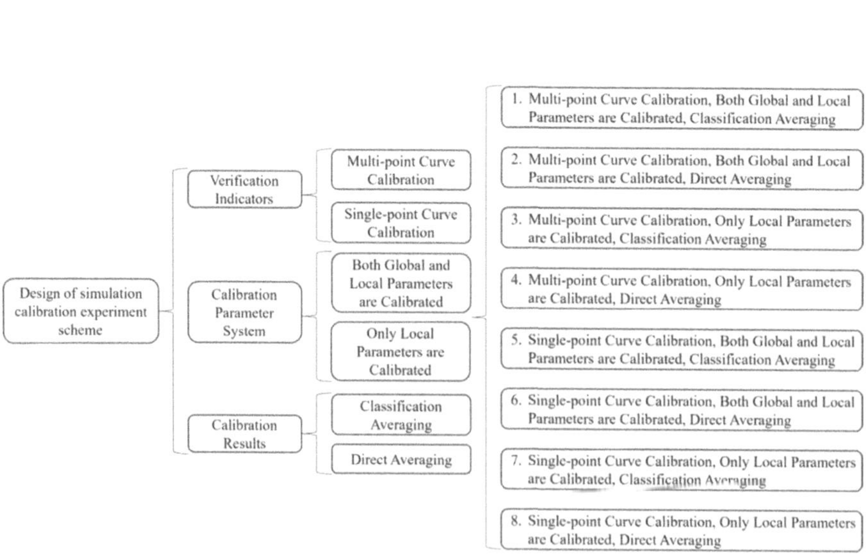

FIGURE 8.6 Design of simulation calibration experiment scheme.

TABLE 8.3
Calibration values of parameters of each experimental scheme

Scheme	Safety distance conversion coefficient	Additional part of safety distance	Multiplicative part of safety distance	Average parking spacing
1	0.13	3.57	5.14	1.13
2	0.17	3.62	5.07	1.18
3	0.31	3.56	5.12	1.24
4	0.26	3.71	4.97	1.12
5	0.84	1.32	1.74	1.74
6	0.83	1.10	1.51	1.50
7	0.79	5.08	5.26	2.34
8	0.81	4.91	5.34	2.13
Default	0.60	2.00	3.00	2.00

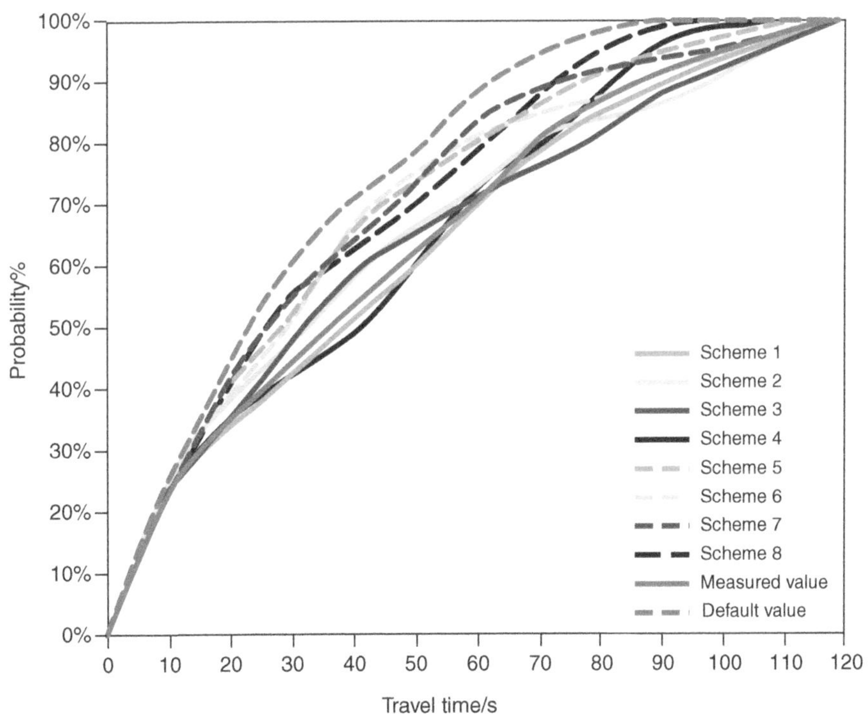

FIGURE 8.7 Cumulative probability distribution curve of simulation output travel time for different experimental schemes.

TABLE 8.4
Similarity and p-value of different experimental schemes

Experimental scheme	Similarity	P-value
Scheme 1	0.156	0.983
Scheme 2	0.236	0.961
Scheme 3	0.228	0.969
Scheme 4	0.214	0.976
Scheme 5	0.659	0.891
Scheme 6	0.634	0.903
Scheme 7	0.86127	0.826
Scheme 8	0.849	0.873
Default	1.124	0.568

approach including multi-curve calibration as a validation measure, global and local parameter classification calibration, and averaging of calibration result classifications.

8.2 MIXED TRAFFIC FLOW SIMULATION

8.2.1 Reinforcement Learning-Based Modeling of Following Control

This section introduces a reinforcement learning-based vehicle following a model based on bilateral control strategy, which can effectively reduce the interference of traffic oscillations. Combining the characteristics of bilateral control strategies, two reinforcement learning-based vehicle following models are proposed. These two models are BCM-RL based on bilateral control strategy and CFM-RL based on unilateral control strategy, which are both trained using deep reinforcement learning.

The state space of CFM-RL is shown in the following equation:

$$s_t = \{v_n(t), d_n(t), \Delta v_n(t)\} \tag{8.6}$$

where $v_n(t)$ is the velocity of a vehicle with the vehicle n at time $t, m/s$; $d_n(t)$ is the distance between the current vehicle of vehicle n and the lead vehicle at time t, m; $\Delta v_n(t)$ is the velocity difference value between vehicle n and preceding vehicle, m/s.

The state space of BCM-RL additionally adds the speed of the rear vehicle and the distance from the front end of the rear vehicle to the rear end of the current vehicle compared to CFM-RL. The state space of BCM-RL is shown in equation 8.9:

$$s_t = \{v_n(t), d_n(t), \Delta v_n(t), v_{n+1}(t), d_{n+1}(t)\} \tag{8.7}$$

In the field of reinforcement learning, the action space is the set of all possible actions that an agent can take at a given time step. The action space in this study refers to the continuous variable of longitudinal acceleration of the self-driving vehicle. To ensure the comfort of the driving process and to make it closer to the driving behavior in the real environment, the range of the action space was limited to [–3 m/s², 3 m/s²] according to Zhu et al. (2020).

The main purpose of the base reward is to allow the model to quickly learn the following behavior that can take into account the optimization goals of safety, driving efficiency, and vibration suppression. For the CFM-RL control efficiency-related term, it is shown below:

$$f_{cfm-control} = (c_n)^T Q_c c_n \tag{8.8}$$

$$Q_{cfm} = \begin{bmatrix} \alpha_{cfm-d} & \\ & \alpha_{cfm-v} \end{bmatrix} \tag{8.9}$$

$$c_n = [v_n(t)^* t_1 + d_1 - d_n(t), \Delta v_n(t)] \tag{8.10}$$

where Q_{cfm} is the positive definite diagonal coefficient matrix; t_1 is the desired inter-vehicle time distance of CFM-RL, s; d_1 is the minimum stopping distance of CFM-RL, m; α_{cfm-d} is the position control coefficient of CFM-RL; and α_{cfm-v} is the speed control coefficient of CFM-RL.

For BCM-RL, this study draws on the core idea of the bilateral control model proposed by Wang et al. (2019). The objectives of this strategy are to ensure that the vehicle is able to travel at an average speed that follows the rear vehicle and leads the front vehicle; and the vehicle to be located in the middle of the rear vehicle and the front vehicle. The terms related to the control efficiency of the bilateral control BCM-RL are shown below:

$$f_{bcm-control} = (b_n)^T Q b_n \tag{8.11}$$

$$Q_{bcm} = \begin{bmatrix} \alpha_{bcm-d} & \\ & \alpha_{bcm-v} \end{bmatrix} \tag{8.12}$$

$$b_n = [d_n(t) - d_{n-1}(t), v_{n-1}(t) - v_{n+1}(t)] \tag{8.13}$$

where Q_b is the matrix of positive definite diagonal coefficients; α_{bcm-d} is the position control coefficients of BCM-RL; and α_{bcm-v} is the velocity control coefficients of BCM-RL.

In terms of suppressing traffic oscillations, it is necessary to constrain the velocity change of the vehicle itself, that is, try to make the self-driving vehicle select acceleration closer to zero from the action space. The BCM-RL term related to the suppression of traffic oscillations, which consists of the quadratic of the acceleration and the absolute value of the acceleration, is shown in equation (8.14), where α_{acc1} and α_{acc2} are positive values, and the model itself is affected by adjusting α_{acc1} and α_{acc2} to influence the acceleration selection.

$$f_{acc} = \alpha_{acc1}(a_n(t))^2 + \alpha_{acc2} \mid a_n(t) \mid \tag{8.14}$$

where α_{acc1} is the quadratic square penalty parameter of the acceleration of the autonomous vehicle; α_{acc2} is the absolute value penalty parameter of the acceleration of the autonomous vehicle.

In terms of reward function design, this study utilizes a logarithmic function that combines terms related to control efficiency and suppression of traffic oscillations to construct a smooth reward function that supports the agent to learn more easily. The CFM-RL reward function is shown in equation 8.15:

$$r_{cfm-base} = -0.1\log(1 + f_{cfm-control} + f_{acc}) \tag{8.15}$$

The reward function of BCM-RL is shown in equation 8.16:

$$r_{bcm-base} = -0.1\log(1 + f_{bcm-control} + f_{acc}) \tag{8.16}$$

8.2.2 Training Environment

This study proposes a nested two-stage training framework to train CFM-RL and BCM-RL separately. Figure 8.8 illustrates the idea of the two-stage training framework adopted in this study. The core objective of the first stage of training is to train the CFM-RL that satisfies the optimization objective and constraints, thus providing effective assistance for the training of the BCM-RL in the second stage. In the second stage, multiple BCM-RLs are trained simultaneously to improve the generalization ability and performance of the model.

During the model training process, a speed profile is randomly selected from the NGSIM dataset in each round, and the human-driven vehicle drives according to this speed profile. At the same time, the initial speed of CFM-RL is consistent with the initial speed of the human-driven vehicle, the vehicle spacing is set to $v_0(0) * T_d + d_0$, and the reinforcement learning algorithm controls the speed of the subsequent vehicles. In this process, the vehicle will adjust its own speed according to the current state and control strategy, and through continuous iteration and optimization, the model will eventually learn a longitudinal control strategy that satisfies multiple optimization objectives in a complex traffic environment (see Figure 8.9).

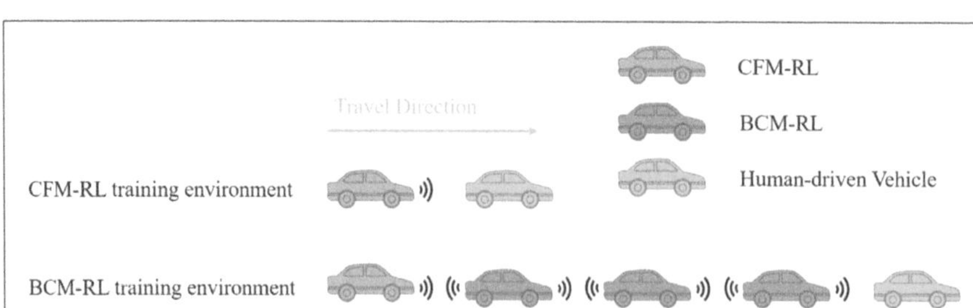

FIGURE 8.8 Schematic diagram of a nested two-stage training framework.

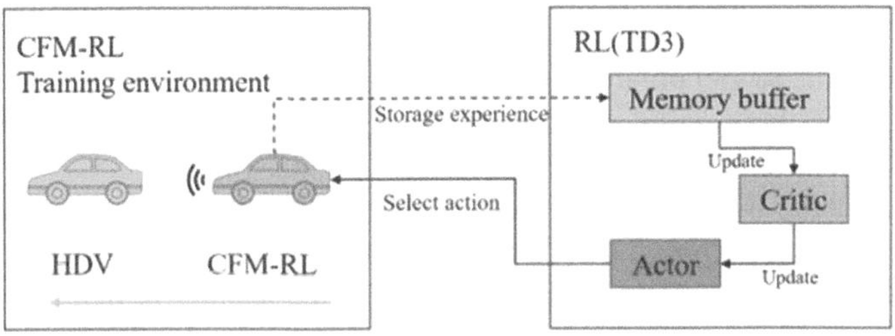

FIGURE 8.9 CFM-RL training framework.

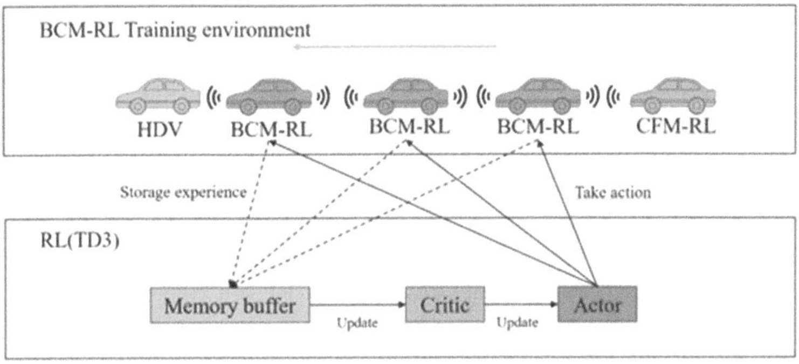

FIGURE 8.10 BCM-RL training framework.

In the Phase II training environment, five vehicles were involved, including the lead vehicle (human-driven vehicle), the CFM-RL vehicle (tail vehicle) obtained from Phase I training, and three BCM-RL vehicles. Echoing the approach of the preliminary training phase, the velocity data for the human-operated lead vehicle continue to be sourced from the established training dataset. In the second phase, the CFM-RL no longer performs update learning, and its main role is to act as the tail vehicle of the self-driving vehicle chain to assist in the training of the BCM-RL. It is worth noting that during this training process, all BCM-RL vehicles share the parameters of one TD3 model as a way to enhance the robustness and stability of the model. As shown in Figure 8.10, all BCM-RL vehicles are selected by the same Actor network for the corresponding actions, while the experience trajectories of each vehicle are stored in the experience playback pool, contributing equally to the Critic network update.

The BCM-RL vehicle initialization operation is as follows:

For vehicles with number i: the initial speed is $v_i(0) = v_0(0)$, where $v_0(0)$ is the initial speed of the human-driven vehicle (lead vehicle). For vehicles with number i, the initial spacing with the front vehicle is $d_i(0) = v_i(0) * T_d + d_0$.

8.2.3 Parameter Settings

To ensure that the model obtains better performance and achieves fast and stable convergence during the training process, this study selects appropriate hyperparameter values through several training attempts.

In terms of the neural network design of the TD3 algorithm, this study did not use images as state inputs but selected a small amount of important information (e.g., speed of the ego vehicle, speed difference to the leading vehicle) as state inputs during the traveling process. Therefore, deep neural networks (DNNs) were used in this study to construct the Actor network and Critic network. In the TD3 algorithm used in CFM-RL, both the two Actor networks and the four Critic networks have two hidden layers with 32 neurons in each layer, while using relu as the activation function. On the output layer, the two Actor networks use the tanh activation function. The tanh function is able to map real values to the range [−1,1]. Therefore, it can be used as the output action by multiplying the output value by 3, that is, the acceleration is limited to between −3 m/s^2 and 3 m/s^2, which meets the practical needs of the real model, and the network structure is shown in Figure 8.11.

Compared to CFM-RL, BCM-RL is more difficult to train. To cope with this, the TD3 algorithm used in BCM-RL increases the number of neurons in the two hidden layers of the Actor network and Critic network, using 64 neurons in each layer. Besides that, the other network structures of BCM-RL are consistent with CFM-RL, which is shown in Figure 8.12.

During model training, Ornstein–Uhlenbeck noise (OU-Noise), which performs well in inertial systems, was chosen as the action noise. OU-Noise simulates the velocity of Brownian particles subjected to friction and generates time-dependent values centered at zero. This time-dependent noise allows intelligent bodies in physical environments with momentum to perform effective exploration. The OU-Noise has a mean of 0.15 and a variance of 0.2 in training. The strategy noise uses normally distributed noise, set to have a mean of 0 and a standard deviation of 0.3. For details on the values of other hyperparameters of the TD3 algorithm used for both models, see Table 8.5.

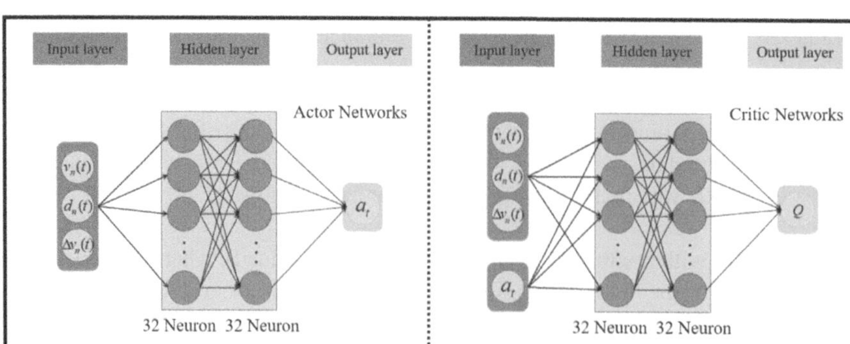

FIGURE 8.11 Actor–Critic network Structure in CFM-RL Model.

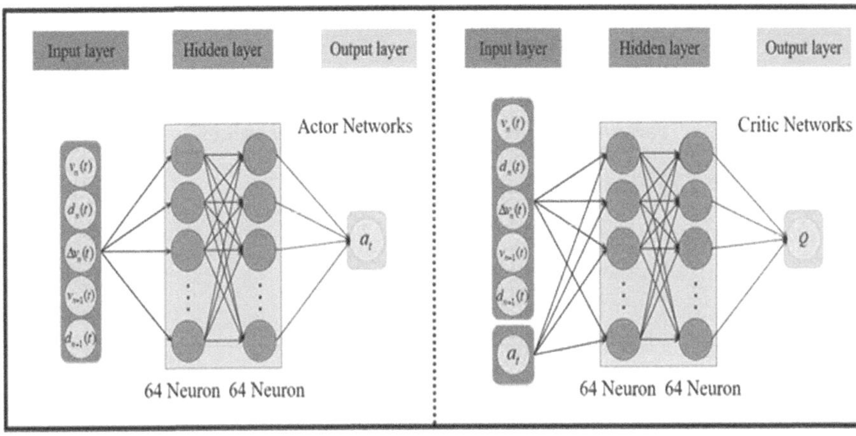

FIGURE 8.12 Schematic diagram of Actor and Critic network structure.

TABLE 8.5
TD3 algorithm parameter settings

Parameter name	CFM-RL parameter values	BCM-RL parameter values
Discount factor	0.95	0.95
Actor network learning rate	0.0001	0.0001
Critic network learning rate	0.0002	0.0002
Types of optimization algorithms	Adam	Adam
Batch size	64	256
Soft update rate	0.001	0.001
Experience pool size	20,000	80,000
Strategy frequency	2	2

In the parameter setting of CFM-RL desired inter-vehicle time distance and minimum stopping spacing aspects are selected as 1.6 seconds and 2 m, respectively. It should be emphasized that the selection of parameters in the reward function has a significant effect on the model performance. In the selection process of hyperparameters of the reward function, this study set the unilateral control model with reference to the study of Zhou et al. (2021), and other parameters were obtained by adjusting through several experiments. The model parameters in this study are shown in Table 8.6.

8.2.4 Mixed Flow Simulation Validation Experiments

In this study, four different scenarios were designed for simulation testing, aiming to examine the ability of BCM-RL vehicle chains of different lengths to improve traffic conditions in mixed traffic flows. In each of these four scenarios, 31 vehicles

TABLE 8.6
Model parameter setting table

Parameter name	Parameter values
α_{cfm-d}	0.4
α_{cfm-v}	0.2
α_{bcm-d}	0.3
α_{bcm-v}	0.5
α_{acc1}	1
α_{acc2}	1

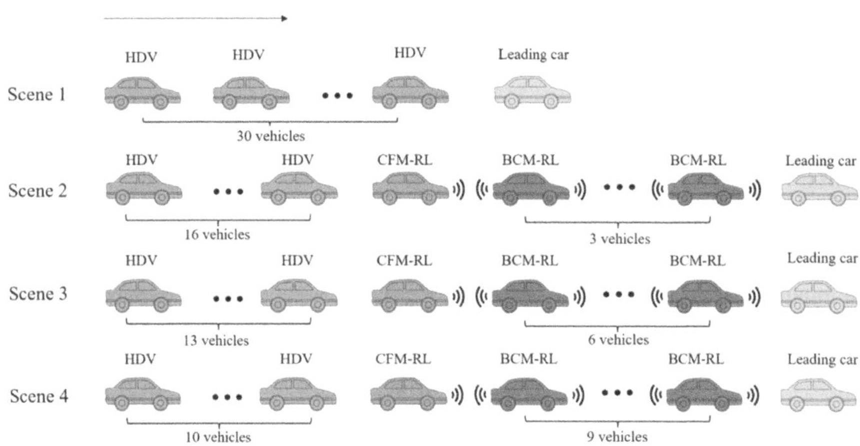

FIGURE 8.13 Mixed traffic test environment.

are included, where the speed data of the lead vehicle is from the test set, and the composition of the 30 following vehicles is shown in Figure 8.13, where the human vehicles use the intelligent driver model (IDM) and the lead vehicle is spliced from the NGSIM vehicle speed profile data.

Figure 8.14 demonstrates the changes in the position of each vehicle over time for the four scenarios. As can be seen from Figure 8.14, in the purely human driving environment of Scenario 1, the speed of the vehicles in the rear part shows an obvious fluctuation due to the influence of traffic waves. However, as the number of BCM-RL vehicles increases, the fluctuation of the time-position curves of the trailing vehicles gradually decreases. Especially in Scenario 4, the position-time curve fluctuations of the trailing vehicles decreased significantly. This result indicates that the BCM-RL vehicle chain can achieve the effects of stabilizing vehicle speed, absorbing traffic waves, and mitigating traffic oscillations, and the longer the BCM-RL vehicle chain, the better the effect.

Figure 8.15 illustrates the variation of each vehicle's speed over time in the four scenarios. In the purely human driving environment of Scenario 1, the speed fluctuations of the trailing vehicle become smaller in magnitude relative to the leading

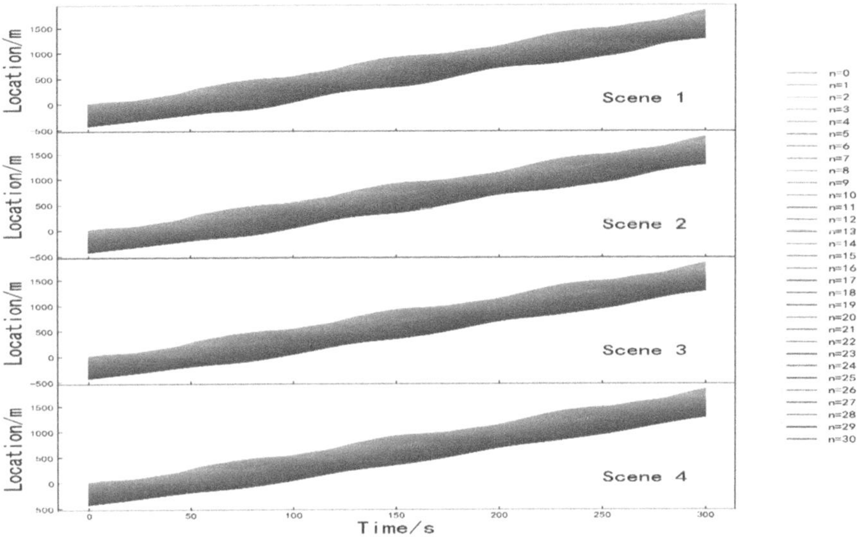

FIGURE 8.14 Changes in the position of each vehicle over time for four scenarios.

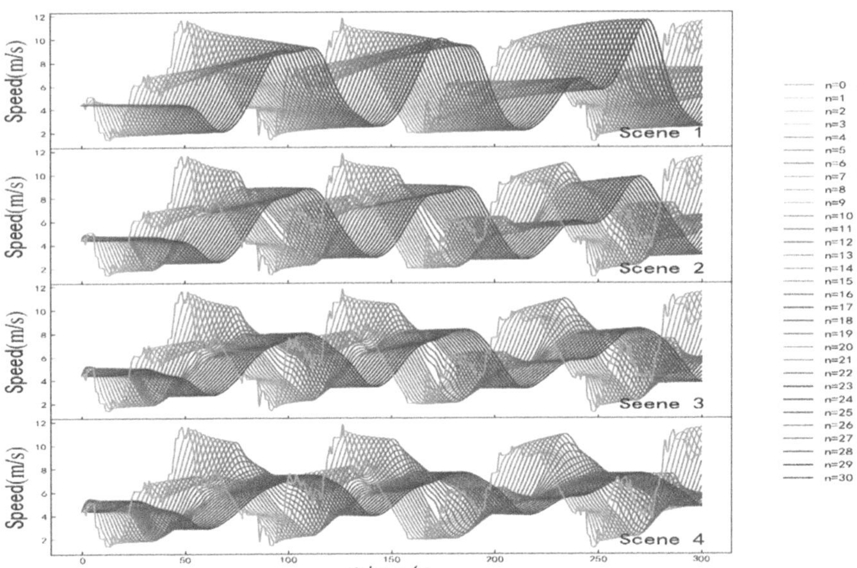

FIGURE 8.15 Change in speed over time for each vehicle in four scenarios.

vehicle from 0s to 200s, indicating that the IDM model can suppress the propagation of oscillations to a certain extent; however, the speed fluctuations of the trailing vehicle are significantly larger than those of the leading vehicle from 250s to 300s, presenting the phenomenon of amplified speed oscillations. In Scenarios 2 to 4, on the other hand, when the speed fluctuations are transmitted to the bilateral control BCM-RL, the fluctuations gradually diminish, and the speed of the trailing vehicle becomes more stable with the growth of the BCM-RL vehicle chain. This indicates that the BCM-RL vehicle chain plays the role of a "stabilizer" in the traffic flow, which can effectively absorb traffic waves and reduce traffic oscillations.

8.3 CONCLUSION AND APPLICATIONS

In recent years, the rapid development of CAV technologies has given rise to smart connected transportation, contributing to the improvement of road traffic efficiency, enhancement of road traffic safety, and conservation of energy consumption. The National Highway Traffic Safety Administration (NHTSA) in the U.S. mandated in 2014 that newly introduced vehicles must possess connected communication capabilities. The Intelligent Transportation Systems Plan (2020–2025) released in the U.S. specifically emphasizes research on intelligent connected traffic control technologies, leading global cooperation and innovative research in intelligent transportation systems. China's "Outline for Building a Strong Transportation Nation" released in 2019 explicitly states the need to strengthen the research and development of connected vehicles (smart vehicles, autonomous driving, vehicle-to-infrastructure cooperation) to form a complete and controllable industrial chain and to develop a new generation of intelligent traffic control systems.

The rapid development of these technologies is driving the transformation of road traffic control objects, data environments, and control methods: traditional traffic flow is evolving into intelligent connected new hybrid traffic flow. The intelligent connected hybrid traffic flow, composed of conventional vehicles (RV), connected vehicles (CV), and CAVs, has already emerged and is expected to persist in the long term. The application of microscopic models in the study of hybrid traffic flow can provide effective insights for future research.

Microscopic simulation tool parameter calibration involves optimizing the definition of global and local parameters through a multi-point curve calibration approach, facilitating the calibration of microscopic simulation software. Future research could build upon this calibration method to investigate rapid parameter calibration methods for mixed traffic flow, aligning simulation environments with the real world to enhance the exploration of mixed traffic flow characteristics.

A car-following control model for mixed traffic flow based on reinforcement learning adopts a bilateral control strategy, considering information from surrounding vehicles. This model not only reduces traffic oscillations but also lowers fuel consumption and tailpipe emissions. Future research could further explore the impact of improved performance of autonomous vehicles on the overall stability of traffic flow in mixed traffic environments.

REFERENCES

Lin, Z., Zhou, C., Rong, J., Li, W., & Li, Q. (2019). An analysis of acceleration-speed parameter threshold based on actual engineering measurement. *Journal of Transport Information and Safety*, (02), 107–113.

Shi, H., Zhou, Y., Wu, K., Wang, X., Lin, Y., & Ran, B. (2021). Connected automated vehicle cooperative control with a deep reinforcement learning approach in a mixed traffic environment. *Transportation Research Part C: Emerging Technologies*, 133, 103421.

Wang, L., & Horn, B. K. (2019). On the chain stability of bilateral control model. *IEEE Transactions on Automatic Control*, 65(8), 3397–3408.

Zhu, M., Wang, Y., Pu, Z., Hu, J., Wang, X., & Ke, R. (2020). Safe, efficient, and comfortable velocity control based on reinforcement learning for autonomous driving. *Transportation Research Part C: Emerging Technologies*, 117, 102662.

9 Case Study

Mixed Traffic with Connected and Automated Vehicle Scenario in Southern California

Brian Yueshuai He, Qinhua Jiang, and Jiaqi Ma

9.1 INTRODUCTION

As a transformative technology poised to revolutionize transportation systems, CAVs have garnered substantial attention in recent years (Fagnant and Kockelman, 2015). This innovative technology holds the potential to bring multifaceted benefits to transportation systems, encompassing safety improvements, economic advantages, societal enhancements, increased efficiency and convenience, and expanded mobility options (NHTSA, n.d.; Ma et al., 2017; Wang et al., 2019). Notably, the elimination of human errors, responsible for 94% of accidents, stands as a pivotal safety advancement. Moreover, CAVs, through in-vehicle equipment and infrastructure connectivity, reduce drivers' perception–reaction time and minimize inter-vehicle distances both laterally and longitudinally. Applications such as platooning can mitigate congestion, subsequently enhancing fuel economy. The accessibility gains provided by CAVs, particularly benefiting younger, older, and disabled individuals, underscore its societal significance.

On the demand side, the advent of CAVs is anticipated to induce travel demand by simplifying travel processes (He et al., 2022), while on the supply side, it holds the potential to enhance roadway capacity and diminish accident rates (Jiang et al., 2022). Addressing the need for empirical insights into the extent of CAV impact, we conducted a case study in Southern California, focusing on the evaluation of CAV effects on transportation systems, with particular emphasis on induced travel demand and the dynamics of mixed traffic scenarios.

9.2 METHODOLOGY

This study addresses two key inquiries arising from the widespread adoption of CAVs: (1) the extent to which changes in individuals' travel behaviour can impact

 DOI: 10.1201/9781032674667-9

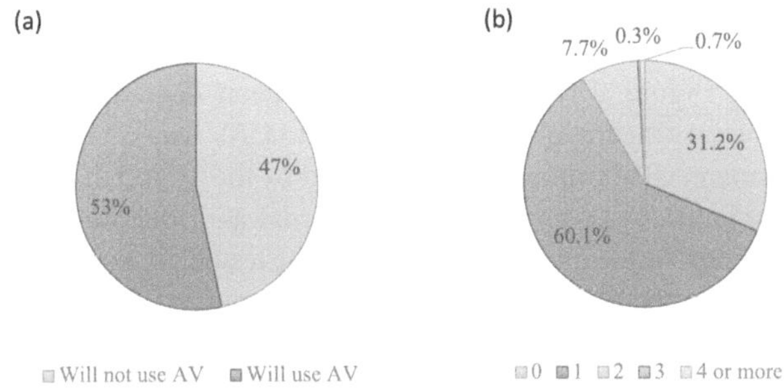

FIGURE 9.1 Methodology framework of case study.

travel demand and (2) the influence of CAVs on transportation systems in mixed traffic scenarios. Our methodological framework, illustrated in Figure 9.1, comprises two main components. On the left side, a survey was meticulously designed to gather information on individuals' inclination to use CAVs and capture behaviour choices associated with CAV utilization. Subsequently, a binary choice model was formulated, leveraging survey data, to determine whether individuals within the synthetic population would opt for CAVs in their daily travel routines. For identified CAV users, their activities and travel demand were derived from the CAV-incorporated Activity-Based Model (ABM), while non-CAV users' data were sourced from the traditional ABM developed by the Southern California Association of Governments (SCAG). The new travel demand encompassed both CAV and non-CAV users. On the right side, the consolidated travel demand was input into the traffic assignment model to estimate traffic conditions in mixed-traffic scenarios. Details of each component will be introduced in subsequent sections.

9.2.1 Survey

We designed a Stated Preference (SP) survey and administered it online to Southern California residents through collaboration with Dynata, a professional survey sampling and data collection company (formerly known as Survey Sampling International or SSI). The survey protocol comprised four distinct steps: (1) Survey programming and quota preparation, where we translated the survey manuscript into the SurveyMonkey platform, simultaneously selecting demographic characteristics as control variables and establishing quotas. (2) Formulating the sampling plan, wherein we determined a sample size of approximately 1000, considering the total budget cap and project timeline. (3) A soft launch involving the delivery of the survey to a small subset (around 10% or 100 respondents) by Dynata. Subsequently, Dynata collected responses and assessed deviations between the soft launch quotas and the target quotas. If substantial biases were identified, the subsequent sampling stage would be executed with more targeted precision. (4) The full survey launch followed issue resolution from the soft launch and consisted of two stages: sampling in the general population to collect socio-economic and demographic information and sampling in CAV users to collect travel-related information.

9.2.2 Behavior Model

As delineated in Figure 9.1, a CAV-ABM has been developed to predict travel behaviours and assess the travel demand specific to CAV users. This model, grounded in the contemporary ABM framework established by the SCAG in 2020, extends from long-term considerations (e.g., work arrangements and location choices) to short-term determinations (e.g., mode choices), as illustrated in Figure 9.2. The model consists of seven layers, commencing with the generation of a synthetic population for the SCAG area, incorporating household-level demographic and socio-economic

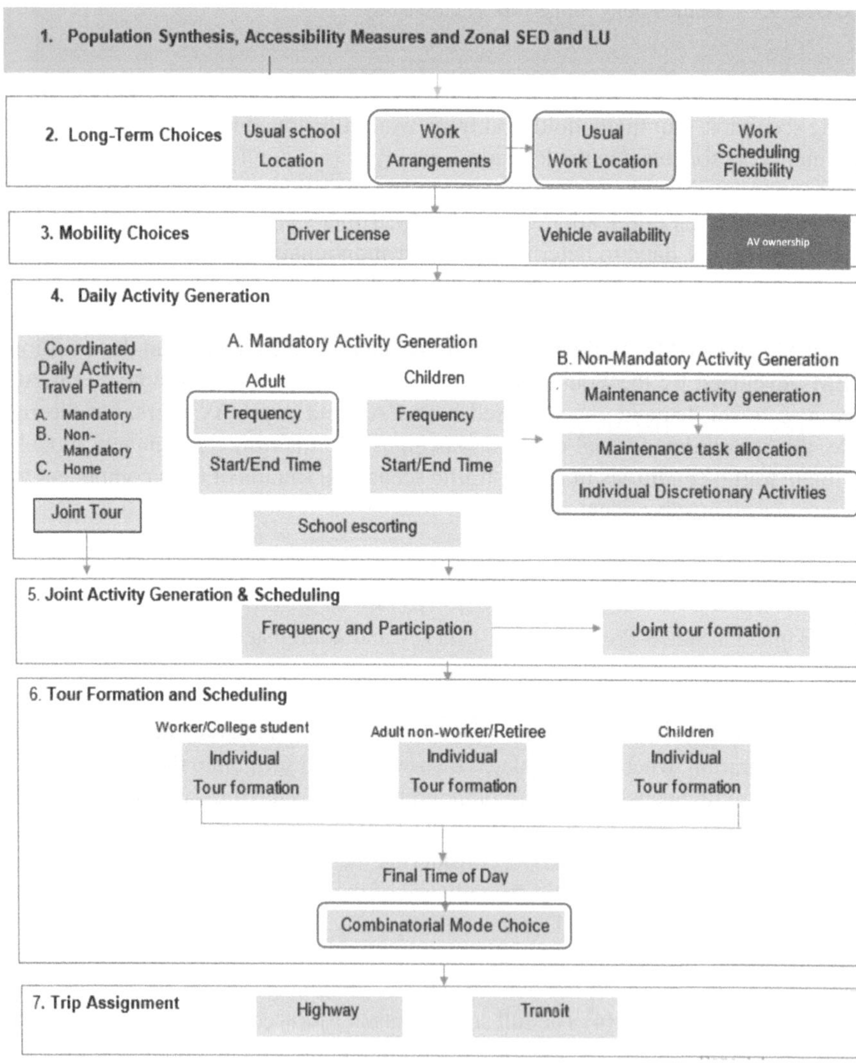

FIGURE 9.2 System design of SCAG ABM (SCAG, 2020) and corresponding updates to incorporate CAV-related behaviour changes (He et al., 2022).

attributes. Subsequent layers involve the generation of activities and tours based on the synthetic person's type (e.g., worker vs. non-worker). The second layer predicts individuals' long-term choices, encompassing factors like work arrangements, usual school and work locations, and work schedule flexibility. Following these long-term decisions, the model proceeds to determine mobility choices in the third layer, such as driver's license possession and vehicle ownership. In the fourth layer, medium-term choices are modelled, considering mandatory activities (e.g., work and school) and non-mandatory activities (e.g., household maintenance and individual discretionary). Temporal constraints on activity windows are also taken into account, with the generation of school escort activities preceding non-mandatory activities in households with school-age children. The fifth layer simulates interactions among household members, while the sixth layer, with all activities generated, formulates tours to link these activities, concurrently determining departure times and mode choices. Ultimately, this activity-based approach culminates in the generation of activity-based travel demand. In the final layer, the SCAG ABM allocates this demand to the physical network (road and transit) to estimate VMT and other performance metrics, contributing valuable insights to inform transportation planning decision-making.

Based on the survey data, a series of models (highlighted by red boxes in Figure 9.2) underwent recalibration to integrate the choice behaviour changes associated with CAV users into the SCAG ABM framework. Specifically, adjustments were made in several layers to accommodate these changes. In the long-term choice layer, recalibrations were implemented to reflect alterations in work arrangement and usual work location preferences among CAV users. Additionally, within the mobility choice layer, a model predicting CAV ownership was developed to estimate the household-level ownership of CAVs, serving as a key input for subsequent layers. Within the fourth layer, updates were applied to the adult mandatory activity frequency model to capture the revised activity patterns observed among individuals utilizing CAVs. Similar modifications were extended to the models governing maintenance and individual discretionary activities. Lastly, the mode choice model underwent revisions to encompass both Personal Automated Vehicles (PAVs) and Shared Automated Vehicles (SAVs) as emerging mode choices. The model estimation results are presented in Tables 9.1–9.3.

In Table 9.1, the coefficients associated with education attainment exhibit a consistent positive trend, with higher education levels yielding larger coefficients, suggesting that individuals with higher levels of education are more inclined to utilize CAV services. Regarding household income, the lowest income bracket (under 25k) serves as the reference level. Notably, coefficients for other income brackets are negative, signifying that individuals with lower incomes exhibit a propensity to utilize CAVs for daily travel. Given the unequal distribution of opportunities among low-income individuals, this demographic is more predisposed to embrace CAV technology to address their underserved travel needs.

Regarding CAV ownership, the coefficients associated with household auto ownership exhibit a positive trend, indicating that as the number of vehicles owned by a household increases, so does the likelihood of purchasing a CAV. This suggests that households possessing a greater number of vehicles are more inclined towards purchasing more

TABLE 9.1
Estimation results of CAV choice model and CAV ownership model

Variable number	Variable name	CAV choice model Coefficient (std)	CAV ownership model Coefficient (std)
	CAV choice constant		
1	Use CAV	0.26 (0.05) ***	
	CAV ownership constant		
2	One CAV		−1.38 (0.13) ***
3	Two CAVs		−3.55 (0.14) ***
4	Three CAVs		−7.58 (0.38) ***
5	Four or more CAVs		−5.75 (0.19) ***
	Age		
6	Age 18 to 24	−0.53 (0.04) ***	
7	Age 25 to 64	−0.54 (0.04) ***	
8	Age 65 +	−0.98 (0.05) ***	
	Education		
9	Associate/2-year college	0.53 (0.02) ***	
10	Bachelor	0.73 (0.02) ***	
11	Master +	0.91 (0.03) ***	
	Gender		
12	Female	−0.14 (0.01) ***	
	Household auto ownership		
13	One		1.85 (0.14) ***
14	Two		2.15 (0.14) ***
15	Three or more		2.84 (0.16) ***
	Household income		
16	25k to 50k		−0.99 (0.08) ***
17	50k to 75k	−0.14 (0.02) ***	0.17 (0.09) *
18	75k to 100k	−0.22 (0.02) ***	1.58 (0.14) ***
19	100k to 125k	−1.46 (0.03) ***	0.83 (0.20) ***
20	125k to 150k	−0.34 (0.03) ***	0.91 (0.16) ***
21	150k +	−0.06 (0.03) *	0.82 (0.11) ***
	Ethnicity		
22	Non-Hispanic White	−0.13 (0.02) ***	
23	Non- Hispanic Black	0.34 (0.03) ***	
24	Non- Hispanic Asian	0.09 (0.02) ***	
25	Non- Hispanic other	0.01 (0.05)	
	Household size		
26	Size = 2	−0.26 (0.02) ***	0.02 (0.09)
27	Size = 3	0.65 (0.02) ***	−0.63 (0.09) ***
28	Size = 4+	0.68 (0.02) ***	−0.22 (0.08) **
	Worker status		
29	Non-worker	0.30 (0.02) ***	

*, **, and *** indicate statistical significance at the 0.05, 0.01, 0.001 levels, respectively

Summary statistics of CAV choice
Number of estimated parameters: 21
Sample size: 671
Initial log likelihood: −68305.44
Final log likelihood: −62953.99
Rho-square: 0.0783

Summary statistics of CAV ownership
Number of estimated parameters: 16
Sample size: 699
Init log likelihood: −11142.32
Final log likelihood: −5437.606
Rho-square: 0.512

Source: He et al. (2022).

TABLE 9.2
Calibration results of work arrangement model and a lookup table of work arrangement alternatives

ID	Variable name	Coefficient (std)	Alternative lookup table			
	Work arrangement constant		**IID**	**Work duration**	**Workplace type**	**Number of jobs**
1	Choice 2	−2.92 (0.26) ***	1	Short	Fixed	One
2	Choice 3	−2.67 (0.24) ***	2	Short	Fixed	Multiple
3	Choice 4	−2.41 (0.23) ***	3	Short	Home	One
4	Choice 5	−1.39 (0.20) ***	4	Short	Home	Multiple
5	Choice 6	−3.68 (0.32) ***	5	Short	Variable	One
6	Choice 7	−1.23 (0.20) ***	6	Short	Variable	Multiple
7	Choice 8	−2.59 (0.24) ***	7	Medium	Fixed	One
8	Choice 9	−2.03 (0.21) ***	8	Medium	Fixed	Multiple
9	Choice 10	−3.88 (0.35) ***	9	Medium	Home	One
10	Choice 11	−2.85 (0.25) ***	10	Medium	Home	Multiple
11	Choice 12	−4.13 (0.38) ***	11	Medium	Variable	One
12	Choice 13	1.02 (0.18) ***	12	Medium	Variable	Multiple
13	Choice 14	−0.63 (0.19) ***	13	Long	Fixed	One
14	Choice 15	−0.63 (0.19) ***	14	Long	Fixed	Multiple
15	Choice 16	−3.44 (0.30) ***	15	Long	Home	One
16	Choice 17	−1.45 (0.20) ***	16	Long	Home	Multiple
17	Choice 18	−1.12 (0.20) ***	17	Long	Variable	One
	Gender		18	Long	Variable	Multiple
18	Female	−0.62 (0.14) ***				
	Household income					
19	25k to 75k	1.15 (0.16) ***				
20	75k to 100k	2.37 (0.25) ***				
21	100k to 125k	2.04 (0.43) ***				
22	125k +	5.25 (0.61) ***				
	Ethnicity					
23	Non-Hispanic White	−1.51 (0.16) ***				
24	Non- Hispanic other	0.98 (0.30) **				

*, **, and *** indicate statistical significance at the 0.05, 0.01, 0.001 levels, respectively

Summary statistics

Number of estimated parameters: 24

Sample size: 395

Init log likelihood: −9665.172

Final log likelihood: −6375.645

Rho-square for the initial model: 0.34

Source: He et al. (2022).

TABLE 9.3
Calibration results and alternative lookup tables of work, maintenance, and discretionary frequency models

Variable number	Variable name	Work frequency	Maintenance frequency	Discretionary frequency
		Coefficient (std)	Coefficient (std)	Coefficient (std)
		Constant		
1	Choice 2	−5.94 (1.02) ***	−2.34 (0.13) ***	−2.34 (0.13) *
2	Choice 3	−0.72 (0.96)	−1.86 (0.12) ***	−1.86 (0.12) ***
3	Choice 4	−2.31 (0.96) *	−2.12 (0.12) ***	−2.12 (0.12) *
4	Choice 5	−6.57 (1.08) ***	−1.55 (0.12) ***	−1.55 (0.12) **
5	Choice 6	−1.89 (0.96) *	−3.54 (0.15) ***	−3.54 (0.15) ***
6	Choice 7	−2.86 (0.96) **	−1.06 (0.12) ***	−1.06 (0.12) **
7	Choice 8	−3.13 (0.96) **	−1.00 (0.12) ***	−1.00 (0.12)
	Age			
8	Age 28 to 24			−0.85 (0.13) ***
9	Age 25 to 64			−1.24 (0.13) ***
10	Age 65+			−0.79 (0.16) ***
		Household auto ownership		
11	One	0.53 (0.33)	−0.85 (0.13) ***	0.83 (0.07) ***
12	Two	3.70 (0.39) ***	−1.24 (0.13) ***	1.65 (0.18) *
13	Three or more	3.70 (0.39) ***	−0.79 (0.16) ***	−0.91 (0.08)
		Household CAV ownership		
14	One	0.99 (0.26) ***	0.83 (0.07) ***	
15	Two or more	0.99 (0.26) ***	1.65 (0.18) ***	1.41 (0.10) ***
		Education		
16	High school or higher	1.79 (0.77) *		
		Gender		
17	Female	0.83 (0.23) ***		−0.57 (0.10) ***
		Household income		
18	25k to 50k	5.47 (0.54) ***	−0.91 (0.08) ***	

TABLE 9.3 (Continued)
Calibration results and alternative lookup tables of work, maintenance, and discretionary frequency models

Variable number	Variable name	Work frequency	Maintenance frequency	Discretionary frequency
		Coefficient (std)	Coefficient (std)	Coefficient (std)
19	50k to 75k	4.33 (0.50) ***	1.41 (0.10) ***	0.22 (0.19) ***
20	75k to 100k	3.57 (0.52) ***	−0.57 (0.10) ***	1.19 (0.10) ***
21	100k to 125k	3.28 (0.57) ***	0.22 (0.19)	0.56 (0.08)
22	125k to 150k	5.85 (0.59) ***	1.19 (0.10) ***	1.08 (0.08)
23	150k+	5.85 (0.59) ***	1.19 (0.10) ***	−2.34 (0.13) ***
		Ethnicity		
24	Non-Hispanic White	−2.11 (0.40) ***		−1.86 (0.12) ***
25	Non- Hispanic Black	−2.11 (0.40) ***		−2.12 (0.12) ***
26	Non- Hispanic Asian	−2.11 (0.40) ***		−1.55 (0.12) ***
27	Non-Hispanic other	−2.11 (0.40) ***		−3.54 (0.15) ***
		Household size		
28	Size = 2 or 3	−3.25 (0.52) ***	0.56 (0.08) ***	
29	Size = 4+	−5.02 (0.54) ***	1.08 (0.08) ***	−1.00 (0.12) ***
		Person type		
30	Part time worker			−0.85 (0.13) ***
31	College student			−1.24 (0.13)
32	Non-worker			−0.79 (0.16) **
33	Retiree			0.83 (0.07) **
34	Driving age children			1.65 (0.18) **
		Worker status		
35	Non-worker			−0.91 (0.08) ***

*, **, and *** indicate statistical significance at the 0.05, 0.01, 0.001 levels, respectively

Summary statistics	Summary statistics	Summary statistics
Number of parameters: 20	Number of parameters: 19	Number of parameters: 32
Sample size: 350	Sample size: 699	Sample size: 699
Init log likelihood: −5624.787	Init log likelihood: −14396.21	Init log likelihood: −14396.21
Final log likelihood: −3429.702	Final log likelihood: −11426.65	Final log likelihood: −9137.41
Rho-square: 0.39	Rho-square: 0.206	Rho-square: 0.365

(continued)

TABLE 9.3 (Continued)
Calibration results and alternative lookup tables of work, maintenance, and discretionary frequency models

Variable number	Variable name		Work frequency		Maintenance frequency		Discretionary frequency	
			Coefficient (std)		Coefficient (std)		Coefficient (std)	
Alt index	Work frequency		Maintenance frequency			Discretionary frequency		
	Work	Business	Shop	Household errands	Escort	Visit	Discretionary	Personal maintenance
1	0	1	0	0	0	0	0	0
2	0	2+	0	0	1+	0	0	1+
3	1	0	0	1+	0	0	1+	0
4	1	1	0	1+	1+	0	1+	1+
5	1	2+	1+	0	0	1+	0	0
6	2	0	1+	0	1+	1+	0	1+
7	2	1	1+	1+	0	1+	1+	0
8	2	2+	1+	1+	1+	1+	1+	1+

Source: He et al. (2022).

CAVs. Additionally, the impact of household income on CAV ownership is noteworthy. Specifically, households with lower incomes (25k–50k) display a negative coefficient, signifying a reduced likelihood of purchasing a CAV. Conversely, as income levels rise, there is a corresponding increase in the propensity to acquire CAVs.

In the long-term phase, we conducted calibrations for both work arrangement, usual work location choices, and number of jobs based on the survey data gathered. The outcomes of these calibrations are outlined in Table 9.2. Additionally, recalibration was performed for short-term activity choices, encompassing work, maintenance, and discretionary frequency models, with the corresponding results detailed in Table 9.3. Further interpretation of these models will be provided in the subsequent section.

9.2.3 Assignment Model

Figure 9.1 illustrates the iterative process and feedback loops between each model component of the traffic assignment model, simulating the mix of traffic of human-driven vehicles and CAVs. The sequence unfolds as follows:

Step 1: Commencing with the demand forecasting phase in the initial loop, travel times are derived from the coded speeds on the input highway networks. Subsequently, the demand model generates Origin–Destination

(OD) trip tables, delineating trips for each vehicle class across various time intervals. These trip tables, segregated by vehicle class and time period, are then assigned to the highway networks, generating initial traffic volumes and congested speeds. Notably, all truck trips are converted to equivalent passenger car trips in the assignment process, employing passenger car equivalent factors identical to those utilized in the SCAG 2016 ABM model (SCAG, 2020), namely 1.2 for light trucks, 1.5 for medium trucks, and 2 for heavy trucks. Additionally, transit vehicles are pre-loaded onto the network, with on-road transits allocated to the highway network and off-road transits assigned to the transit network.

Step 2: The congested speeds obtained from the preceding assignment process are subsequently employed as inputs for the demand model in the subsequent loop. The updated demand yields new OD trip tables, subsequently loaded onto the network links to generate a fresh set of congestion speeds. To mitigate volume discrepancies between consecutive assignment loops, a widely utilized traffic assignment technique known as successive averages is employed. Here, the step size for the averaging method is $1/n$, with n representing the number of iterations, and equilibrium is attained when the relative gap falls below 0.001 or after 200 iterations of the averaging process. It is important to note that the equilibrium referenced here pertains specifically to the convergence of the volume smoothing procedure, not the entire model convergence.

Step 3: The travel times processed through successive averages in preceding loops are then introduced into the final loop, where the last assignment is conducted across all time periods.

To integrate the enhancement of CAVs on roadway capacity within our network, we have adopted highway capacity adjustment factors derived from the research conducted by Adebisi et al. (2020, 2022). Their comprehensive study yielded planning-level capacity adjustment factors tailored for assessing the impact of CAVs on Highway Capacity Manual (HCM) capacities in a future scenario dominated by CAVs. These adjustment factors are contingent upon the CAV market penetration rate (MPR) and the traffic volume of the link under consideration. In our study, we have applied these adjustment factors to all freeway and arterial links in accordance with established HCM guidelines. Furthermore, to approximate the MPR, we have utilized the proportion of CAV-mode trip distance relative to the total trip distance by car-mode, as outlined in equation 9.1.

$$MPR = \frac{D_{CAV}}{\sum_{i \in M} D_i} \tag{9.1}$$

where $M = \{CAV, SOV, HOV\,2driver, HOV\ 3driver, taxi\}, D_i$ refers to the total travel distance of each of the modes in M, and D_{cav} refers to the total travel distance of CAVs.

9.3 RESULTS AND DISCUSSION

9.3.1 Survey Results

The initial valid survey sample comprised 687 respondents in the first stage, with 675 samples collected from CAV users after sample size expansion in the second stage. During the first stage, approximately 53% of individuals expressed a preference for using CAVs for their daily travel, a proportion closely aligned with the findings (52%) of the 2017 California Vehicle Survey (California Energy Commission, 2017), as depicted in Figure 9.3(a). Additionally, respondents' inclination towards purchasing CAVs was assessed during the initial stage, with the results illustrated in Figure 9.3(b).

As depicted in Figure 9.4(a), individuals under the age of 39 constitute a higher proportion of CAV users, whereas those aged 40 and above are more inclined to be non-CAV users. Regarding the distribution of annual household income (Figure 9.4(b)), a greater proportion of CAV users are observed within income brackets ranging from 25k to 75k and 125k to 200k. Notably, the most significant disparity emerges within the 100k to 125k income range, where individuals are more inclined to be non-CAV users. Additionally, concerning education attainment (Figure 9.4(c)), individuals with an associate degree and above exhibit a higher likelihood of using CAVs.

9.3.2 Changes in Travel Behaviour

The calibrated CAV-ABM reveals substantial shifts in travel behaviours. Regarding work arrangements, as illustrated in Figure 9.5(a), there is a notable decrease in the proportion of individuals opting for type 13 work arrangements, characterized by long work durations, fixed workplaces, and a single job, declining from 58% to 45%. Conversely, preferences shifted towards types 5, 15, and 18 arrangements. Type 5 denotes short work durations, variable workplaces, and a single job; type 15 represents long work durations with a home workplace and a single job; and type 18

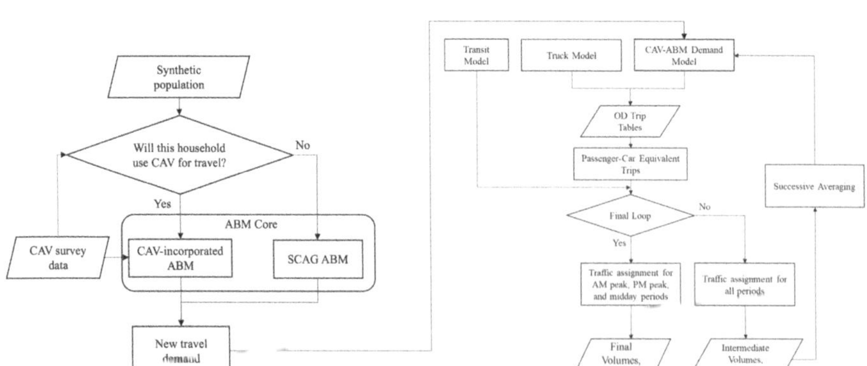

FIGURE 9.3 (a) Distribution of people's intent to use CAV. (b) Distribution of the number of CAVs per household (He et al., 2022).

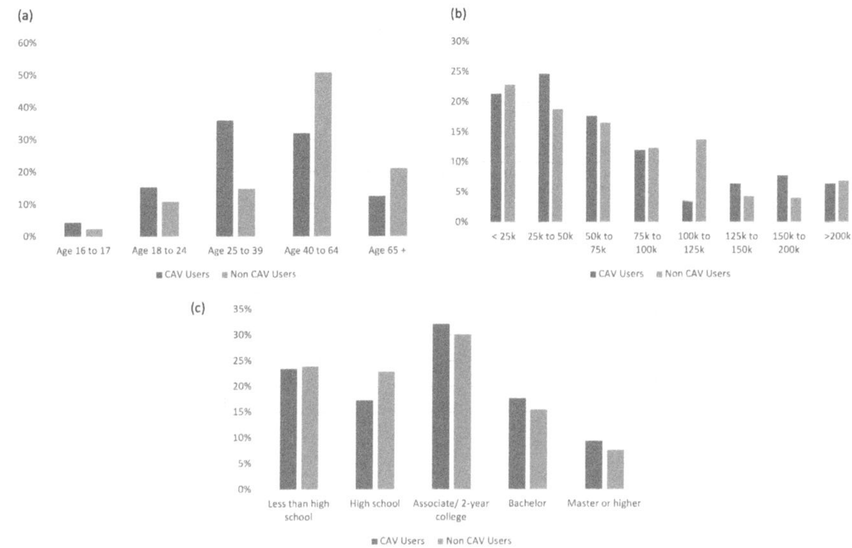

FIGURE 9.4 Distributions of (a) age, (b) annual household income, (c) education attainment of CAV users and non-CAV users (He et al., 2022).

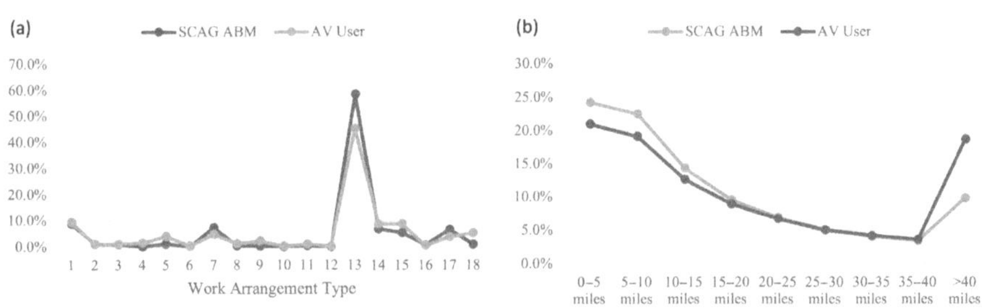

FIGURE 9.5 Comparison of (a) work arrangement predictions and (b) work–home distance from SCAG and CAV-incorporated ABM (He et al., 2022).

encompasses long work durations with variable workplaces and multiple jobs. With the availability of CAVs, individuals exhibited a preference for more flexible work arrangements, favouring shorter work durations, home-based workplaces, and variable workplaces with multiple jobs. This shift can be attributed to the convenience afforded by CAVs, making commutes more manageable and less stressful for workers, thereby facilitating diverse work styles. Additionally, the distribution of work–home distances experienced considerable alterations, as depicted in Figure 9.5(b). Survey data indicated a willingness to accept longer travel times (averaging 12.2 minutes) due to reduced driving stress and increased work productivity. Notably, many CAV users previously opting for shorter distances (less than 20 miles) shifted toward longer distances (more than 40 miles).

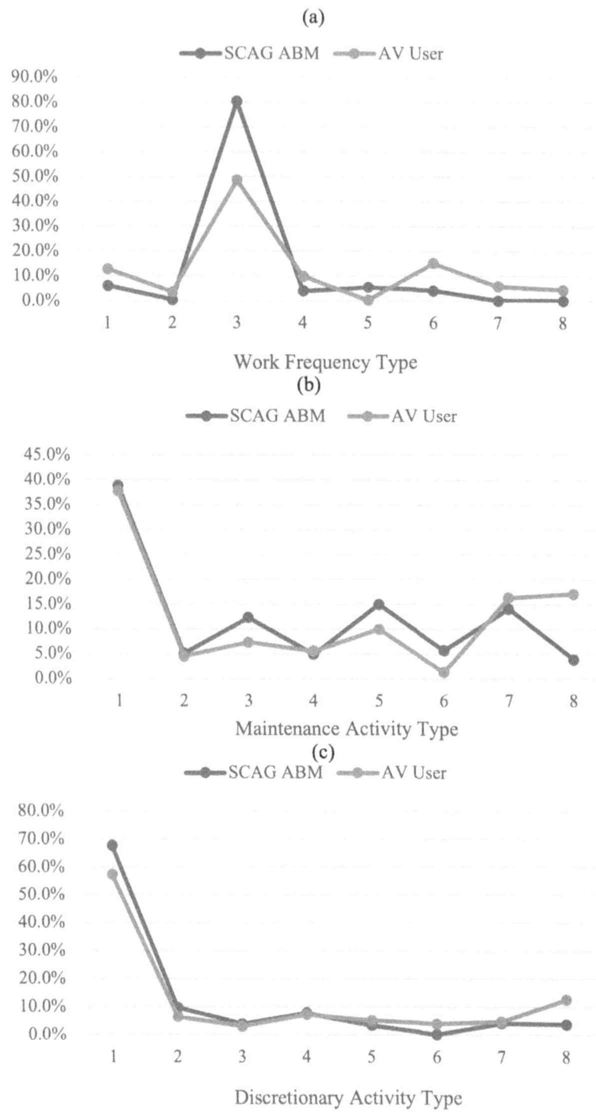

FIGURE 9.6 Distributions of activity frequency (a) work, (b) maintenance, (c) discretionary from SCAG and CAV-incorporated ABM

In the analysis of activity frequency predictions, as depicted in Figure 9.6, noteworthy shifts were observed among CAV users, particularly in work frequency types. Notably, individuals shifted from type 3, characterized by a single work episode, towards types 7 and 8. Type 7 denotes two work episodes and one business episode, while type 8 encompasses two work episodes and two or more business episodes. This trend suggests a propensity among CAV users for increased engagement in both work and business activities, aligning with their preference for flexible work arrangements observed in the work arrangement model. Conversely, changes in maintenance and discretionary frequency were comparatively minor. These activities, predicted subsequent to long-duration mandatory activities, such as work and school, were constrained by limited remaining time windows for allocation.

Table 9.4 illustrates the observed changes in travel behaviours upon the widespread availability of CAVs to the public. The total number of trips saw a notable increase of 7%, with specific categories exhibiting varying degrees of growth. Work-related trips experienced a marked uptick of 14%, while maintenance and discretionary trips increased by 5% and 7%, respectively, and school-related trips remained relatively stable. When considering trips undertaken via car-like modes, the growth rates exhibited a more pronounced increase, with a 15% rise in total car-like mode trips. Notably, work-related trips saw a 16% increase, while school-related trips increased by 7%, and maintenance and discretionary trips witnessed growth rates of 4% and 31%, respectively. This disparity in growth rates suggests that the surge in work, school, and discretionary activities was primarily serviced by car-like modes.

9.3.3 Changes in Traffic Conditions

The simulation outputs from both the SCAG model and the CAV base model are summarized in Table 9.5, revealing notable increases in VMT, the number of trips, and trip distances in the CAV scenario. This surge underscores the induced demand stemming from the widespread deployment of CAVs, resulting in an upsurge in trips and, consequently, elevated miles travelled across the network. In comparison to the SCAG base model, the total VMT witnessed a considerable uptick of 9.1% following the adoption of CAVs, surpassing the growth rate of the number of trips per household (6.8%). This disparity suggests a greater inclination among individuals to embark on longer journeys with the advent of CAVs.

Moreover, Table 9.5 highlights a discernible increase in average freeway speeds during AM peak hours post-CAV adoption, with minimal alterations observed during PM peak hours and on arterials. Given the relatively unchanged trip numbers during AM peak hours due to stringent schedules, the capacity improvements afforded by CAVs translate into enhanced freeway speeds. Conversely, the substantial increase in trips during PM peak hours offsets these capacity enhancements, resulting in marginal alterations in travel speeds. This observation aligns with the marginal increase in Vehicle Hours Travelled (VHT) compared to VMT, as vehicles tend to achieve higher speeds on freeways during AM peak hours. Notably, on arterials, there is only a slight improvement in traffic speeds during AM peak hours, while PM speeds witness a decline in the CAV model. This trend suggests that the induced demand on arterials offsets the capacity enhancements facilitated by CAVs, exacerbating congestion during PM peak hours on arterials.

The spatial distribution of travel speed changes post-adoption of CAVs is depicted in Figures 9.7 and 9.8. Focusing primarily on Los Angeles (LA) County, we utilize the ratio of mean speed to free flow to delineate congestion levels across different time periods. Figures 9.7(a) and 9.8(a) offer a broader overview, while Figures 9.7(b) and 9.8(b) provide zoomed-in perspectives of downtown LA. Figure 9.7 presents the

TABLE 9.4
Changes in total trip numbers and average trip length by car-like mode

		Work	School	Maintenance	Discretionary	Total
Total number of trips	SCAG	20,533,312	17,236,104	15,262,859	18,005,284	71,037,559
	CAV	23,417,304	17,309,718	16,093,094	19,330,363	76,150,479
	Change in %	14%	0%	5%	7%	7%
	Contribution to increase of trip numbers	56%	1%	16%	26%	100%
Total number of trips by car-like modes	SCAG	17,969,234	7,414,271	9,147,706	9,406,251	43,937,462
	CAV	20,877,319	7,944,686	9,474,854	12,344,827	50,641,686
	Change in %	16%	7%	4%	31%	15%
	Contribution to increase of trip numbers	43%	8%	5%	44%	100%
Average trip length (mile)	SCAG	11.69	5.03	7.07	6.88	8.58
	CAV	11.69	4.75	6.48	6.20	8.29
	Change in %	0%	–5%	–8%	–10%	–3%

Source: He et al. (2022).

TABLE 9.5
Comparison between SCAG base model and the CAV scenario

Metrics	Value	SCAG base	CAV
VMT (mi)	Total daily (% change)	4.57×10^8	4.99×10^8 (9.1%)
VHT (h)	Total daily (% change)	1.39×10^7	1.44×10^7 (3.6%)
Trips	Number of trips per household (% change)	12.8	13.67 (6.8%)
Trip length (mi)	Average trip length (% change)	7.18	7.22 (0.61%)
Speed (mph)	Freeway AM peak speed	52.79	55.65 (5.4%)
	Freeway PM peak speed	52.82	53.02 (0.4%)
	Arterial AM peak speed	29.76	29.88 (0.4%)
	Arterial PM peak speed	29.52	29.13 (−1.3%)
Mode share	SOV	40.0%	38.0%
	HOV	48.0%	32.3%
	Transit	1.97%	6.32%
	CAV	NA	14.0%
	Other	10.1%	9.32%
Daily emission (ton)	NOx	26.57	29.22 (9.6%)
	PM2.5	0.67	0.74 (10.4%)
	CO2	1.39×10^{5}	1.53×10^{5} (10.1%)
	ROG	6.61	7.25 (9.7%)

congestion status during the AM period between the SCAG base model and the CAV model. Predominantly, congestion clusters along freeways and arterials near downtown LA, notably I-405, I-10, and US-101, are denoted by the blue circles. With CAV deployment, congestion alleviates notably across multiple freeways surrounding downtown LA, as evidenced by the dash circles in Figure 9.7(b) and 9.7(d).

In contrast, Figure 9.8 offers insights into PM traffic speeds in both the SCAG and CAV models. Compared to the AM period, PM hours exhibit exacerbated traffic congestion, particularly evident in Figure 9.8. Notably, the improvements in traffic speeds with CAV adoption during the PM period are less pronounced compared to the AM period. Fewer freeways and arterials display congestion reduction with CAV utilization in Figure 9.8, indicating a relatively modest impact on congestion mitigation during PM hours.

9.4 CONCLUSION

Recent studies have delved into the potential impacts of CAV technologies on future travel patterns. The advent of CAVs promises to reshape individuals' lifestyles, travel habits, and engagement in various activities. Consequently, comprehending the implications of CAV deployment is crucial for stakeholders to proactively adapt to forthcoming changes.

In this case study, we investigated the impacts of CAVs on people's long-term to short-term travel behaviours by an activity-based approach. This endeavour involved

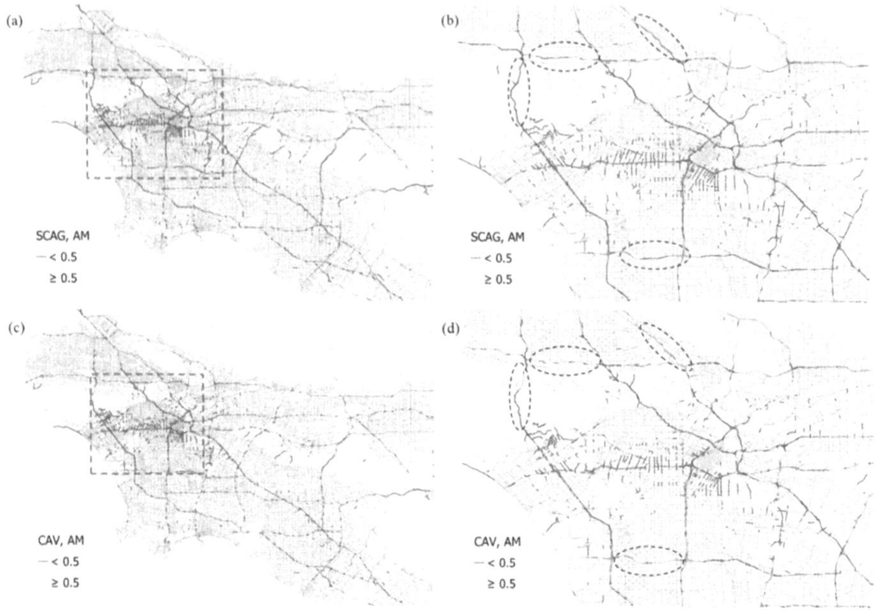

FIGURE 9.7 Ratio of mean speed to free-flow speed during AM peak: (a) SCAG; (b) zoom-in view of (a); (c) CAV; and (d) zoom-in view of (c) (He et al., 2022).

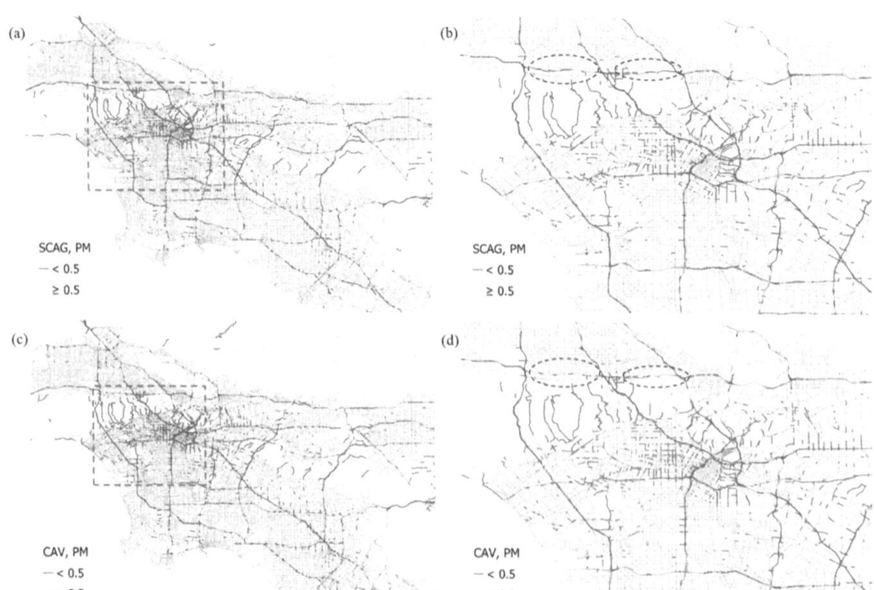

FIGURE 9.8 Ratio of mean speed to free-flow speed during PM peak: (a) SCAG model; (b) zoom-in view of (a); (c) CAV base model; and (d) zoom-in view of (c) (He et al., 2022).

conducting a detailed stated preference survey in Southern California and seamlessly integrating the survey findings with the SCAG ABM. The new travel demand with the deployment of CAV was estimated. Moreover, our study integrated the supply-side impacts of CAVs into traffic assignment modelling, thus incorporating the interplay between travel demand and infrastructure capacity. Through iterative feedback loops, we generated system-level outputs, providing nuanced insights into the dynamic interactions between demand and supply.

Major findings and conclusions from this study are summarized as follows:

- Approximately 53% of individuals in the SCAG area expressed willingness to adopt CAVs for their daily travel. Specifically, individuals with higher education levels (above high school) or belonging to larger households (three or more members) exhibited a greater propensity to embrace CAV technology.
- CAV users exhibited a preference for flexible work arrangements, including remote or variable workplaces with multiple job engagements. Additionally, they displayed a willingness to accept longer work–home distances.
- The total number of trips increased by 7%, with trips via car-like modes experiencing a more substantial uptick of 15%. This suggests a preference for car-like modes amidst increased travel demand, with discretionary trips exhibiting the highest growth rates.
- The VMT and emissions increased significantly with the deployment of CAV. Given the base year model, 2016 SCAG ABM, and a model calibration by survey collected CAV users' data (53% of the population are CAV users), the predicted travel demand could lead to a 9.1% increase in VMT and a 9.6%–10.4% increase in emission.

REFERENCES

Adebisi, A., Guo, Y., Schroeder, B., Ma, J., Cesme, B., Bibeka, A., & Morgan, A. (2022). Highway capacity manual capacity adjustment factor development for connected and automated traffic at signalized intersections. *Journal of Transportation Engineering, Part A: Systems*, 148(3), 04021121.

Adebisi, A., Liu, Y., Schroeder, B., Ma, J., Cesme, B., Jia, A., & Morgan, A. (2020). Developing highway capacity manual capacity adjustment factors for connected and automated traffic on freeway segments. *Transportation Research Record*, 2674(10), 401–415.

California Energy Commission. (2017). California vehicle survey. Retrieved from: www.energy.ca.gov/data-reports/surveys/california-vehicle-survey

Fagnant, D. J., & Kockelman, K., (2015). Preparing a nation for autonomous vehicles: Opportunities, barriers and policy recommendations. *Transportation Research Part A: Policy and Practice*, 77, 167–181.

He, B. Y., Jiang, Q., & Ma, J. (2022). Connected automated vehicle impacts in Southern California part-I: Travel behavior and demand analysis. *Transportation Research Part D: Transport and Environment*, 109, 103329.

Jiang, Q., He, B. Y., & Ma, J. (2022). Connected automated vehicle impacts in Southern California part-II: VMT, emissions, and equity. *Transportation Research Part D: Transport and Environment*, 109, 103381.

Ma, J., Li, X., Zhou, F., Hu, J., & Park, B. B., (2017). Parsimonious shooting heuristic for trajectory design of connected automated traffic part II: Computational issues and optimization. *Transportation Research Part B: Methodological*, 95, 421–441.
NHTSA. (n.d.). Automated vehicles for safety. Retrieve from: www.nhtsa.gov/technology-innovation/automated-vehicles-safety.
SCAG. (2020). *2016 Regional Travel Demand Model and Model Validation Report*. Retrieve from https://scag.ca.gov/sites/main/ les/ le-attachments/validationsummaryreport_20rtp_ nal_2020_05.pdf?1659028273
Wang, J., Peeta, S., & He, X., (2019). Multiclass traffic assignment model for mixed traffic flow of human-driven vehicles and connected and autonomous vehicles. *Transportation Research Part B: Methodological*, 126, 139–168.

Index

Note: Page numbers in **bold** refer to tables.

For Product Safety Concerns and Information please contact our
EU representative GPSR@taylorandfrancis.com Taylor & Francis
Verlag GmbH, Kaufingerstraße 24, 80331 München, Germany